Wireless Body
Area Network

RIVER PUBLISHERS SERIES IN INFORMATION SCIENCE AND TECHNOLOGY

Volume 8

Consulting Series Editors
Prof. KC Chen
National Taiwan University, Taipei
Taiwan

Information science and technology ushers 21st century into an Internet and multi-media era. Multimedia means the theory and application of filtering, coding, estimating, analyzing, detecting and recognizing, synthesizing, classifying, recording, and reproducing signals by digital and/or analog devices or techniques, while the scope of "signal" includes audio, video, speech, image, musical, multimedia, data/content, geophysical, sonar/radar, bio/medical, sensation, etc. Networking suggests transportation of such multimedia contents among nodes in communication and/or computer networks, to facilitate the ultimate Internet. Theory, technologies, protocols and standards, applications/services, practice and implementation of wired/wireless networking are all within the scope of this series. Based on network and communication science, we further extend the scope for 21st century life through the knowledge in robotics, machine learning, cognitive science, pattern recognition, quantum/biological/molecular computation and information processing, biology, ecology, social science and economics, user behaviors and interface, and applications to health and society advance.

- Communication/Computer Networking Technologies and Applications
- Queuing Theory, Optimization, Operation Research, Stochastic Processes, Information Theory, Statistics, and Applications
- Multimedia/Speech/Video Processing, Theory and Applications of Signal Processing
- Computation and Information Processing, Machine Intelligence, Cognitive Science, Decision, and Brain Science
- Network Science and Applications to Biology, Ecology, Social and Economic Science, and e-Commerce

For a list of other books in this series, see final page.

Wireless Body Area Network

Huan-Bang Li

Kamya Yekeh Yazdandoost

Bin Zhen

River Publishers

Aalborg

Published, sold and distributed by:
River Publishers
PO box 1657
Algade 42
9000 Aalborg
Denmark
Tel.: +4536953197

ISBN: 978-87-92329-46-2
© 2010 River Publishers

Preface

Networking technologies have been bringing great changes into communications or connections between people to people, people to multimedia, people to services, and so on. Wireless communication technologies enable ubiquitous networking for anyone, at anytime, and at anywhere, thus can significantly increase convenience and quality of life. Various wireless networks have been developed for different purposes and different application scenarios.

Wireless body area network (WBAN) is a small-scaled network that operates inside, on, or in peripheral proximity of a body. Although communications among devices on a body like wearable computer have been proposed and studied since late 1990s, it is the up-to-date advances in microelectronics and wireless technologies that heighten the technical feasibilities. The strong demands from medical and healthcare society as well as from consumer electronics industry have been driving the beginning of WBAN. The setup of BAN interest group (BAN-IG) within IEEE 802.15 working group (WG15) on wireless personal area network (WPAN) in May 2006 was a significant step that accelerated the WBAN-related research and development activities. Later, the BAN-IG was approved as a task group 6 (TG6) in December 2007 by the IEEE 802 Local and Metropolitan Area Network Standards Committee which fastened the procedure to make an IEEE standard for WBAN. During the same period of time, there were also several important related undertaken. Continua Health Alliance was formed to develop personal healthcare solutions, which is now composed of more than 200 companies worldwide. ETSI Project on eHealth in Europe and Medical ICT Consortium in Japan were set up also for developing wireless and networking technologies to assist healthcare and/or medical applications. In all these industries-oriented collaborations, WBAN is expected to be one of the main technologies of providing extremely high convenience and high efficiency in assisting healthcare or medical services. From the consumer electronics point of view, WBAN is also of great interest in providing body-centric electronics for leisure, entertainment, game control, etc. It is easy to imagine that WBAN can support a huge range of applications. When used for medical and healthcare purposes, WBAN devices will operate with build-in or connected biosensors to collect various vital signals. For body-centric electronics, WBAN devices may also be implemented with action sensors to enable interaction between body and machines. Nevertheless, data streaming or audio/video delivery among body carried machines are also good applications for WBAN.

WBAN has many particular characteristics when compared with other wireless networks. First, besides some common frequency bands, there are various allocated frequency spectrums in different regulations for medical and healthcare purposes. Moreover, electromagnetic wave propagation in WBAN is much different from others since body will be involved as major channel impairment. In most cases, WBAN devices are distributed on or close to body tissues and organs, safety to human body becomes an important issue. Furthermore, to satisfy the requirements of consecutive medical vigilance or healthcare check, WBAN devices need to operate on button or even smaller batteries in long time period. Therefore, extremely low power consumption is of central importance. Thus, power efficient modulation and channel coding as well as power saving medium access control (MAC) are essential in designing a good WBAN system. Coexistence between WBAN and other existing wireless networks also needs to be paid special attention. On one hand, WBAN data cannot be degraded by other wireless systems because of the requirements of high reliability and secure data delivery especially for medical and healthcare-related applications. On the other hand, WBAN must not interfere with other medical systems especially when used in hospitals or clinics.

This book addresses various aspects of WBAN. The objective is to provide sound understanding of the basic concepts, characteristics, and technologies of this new fast growing wireless system. It is the authors' sincere desire that the book can be a useful tool for university students as well as communication system engineers who study or design WBAN.

Acronym

ABC	Absorbing Boundary Conditions
ACK	Acknowledgement
ACS	Add-Compare-Select
AFA	Adaptive Frequency Agility
AFD	Average Fading Duration
APSK	Amplitude-Phase Shift Keying
ARQ	Automatic Repeat reQuest
ASK	Amplitude Shift Keying
AWGN	Additive White Gaussian Noise
BAN	Body Area Network
BER	Bit Error Rate
BPM	Burst Position Modulation
BPSK	Binary PSK
BTMA	Busy tone multiple access
BWA	Broadband Wireless Access
CAP	Contention Access Period
CCA	Clear Channel Assessment
CDF	Cumulative Distribution Function
CDMA	Code Division Multiple Access
CFP	Contention Free Period
CFR	Code of Federal Regulations
CSMA	Carrier Sense Multiple Access
CSS	Chirp Spread Spectrum
DAA	Detect-and-Avoid
DBPSK	Differential BPSK
DPSK	Differential PSK
DSSS	Direct Sequence Spread Spectrum

E	Electric
EAP	Exclusive Access Phase
EC	European Commission
EIRP	Equivalent Isotropic Radiated Power
EM	Electromagnetic
EMC	Electromagnetic Compatibility
EMI	Electromagnetic Interference
ETSI	European Telecom Standards Institute
FBW	Fractional Bandwidth
FCC	Federal Communications Commission
FDMA	Frequency Division Multiple Access
FDTD	Finite Difference Time Domain
FEM	Finite Element Method
FFT	Fast Fourier Transform
FHSS	Frequency Hopping Spread Spectrum
FSK	Frequency Shift Keying
FSMC	Finite-state Markov Chain
GF	Galois Field
GFSK	Gaussian Filtered FSK
GMSK	Gaussian Filtered MSK
GTS	Guaranteed Time Slot
GURL	General User Radio License
IBM	IBM International Business Machines
IEC	International Electrotechnical Commission
IEEE	Institute of Electrical and Electronic Engineers
ICNIRP	International Commission on Non-Ionizing Radiation Protection
IDFT	Inverse Discrete Fourier Transform
IMT	International Mobile Telecommunication
IR	Impulse Radio
ISDB-T	Integrated Services Digital Broadcasting for Terrestrial
ISM	Industrial, Scientific and Medical
ISO	International Organization for Standardization
ITU-R	International Telecommunication Union
ITU-R	International Telecommunication Union Radio Communication Sector
LBT	Listen-Before-Talk
LCR	Level Crossing Rate

LDC	Low Duty Cycle
LIPD	Low Interference Potential Devices
LNA	Low Noise Amplifier
LPF	Low Pass Filter
M	Magnetic
MAC	Medium Access Control
MIC	Ministry of Internal Affairs and Communications
MICS	Medical Implant Communications Service
MID	Medical Implemented Device
MIE	Medical Implant Event
MIIT	Ministry of Industrial and Information Technology
MIT	Massachusetts Institute of Technology
MITS	Medical Implant Telemetry System
MoM	Method of Moment
MSK	Minimum Shift Keying
NFC	Near Field Communication
NLOS	Non-Line of Sight
NOI	Notice of Inquiry
NPRM	Notice of Proposed Rule Making
NRPB	National Radiological Protection Board
NTIA	National Telecommunications and Information Administration
OFDM	Orthogonal Frequency Division Multiplexing
OOK	On-Off Keying
OQPSK	Offset QPSK
OSI	Open System Interconnection
PAP	Priority Access Period
PAPR	Peak-to-Average Power Ratio
PDF	Probability Distribution Function
PHY	Physical Layer
PML	Perfect Metallic Layers
PN	Pseudo Noise
PPM	Pulse Position Modulation
PRF	Pulse Repetition Frequency
PRI	Pulse Repetition Interval
PSD	Power Spectrum Density
PSK	Phase Shift Keying
PTS	Priority Time Slot
PWM	Pulse Width Modulation

QAM	Quadrature Amplitude Modulation
QPSK	Quadrature PSK
RAP	Random Access Phase
RFID	Radio Frequency Identification
RR	Radio Regulation
RF	Radio Frequency
RFID	Radio Frequency Identification
Rx	Receiving Antenna
SA	Specific Absorption
SAR	Specific Absorption Rate
SNR	Signal-to-Noise Ratio
SRCF	Square-root Raised Cosine Filter
SRD	Short Rang Radio
TDMA	Time Division Multiple Access
Tx	Transmitting Antenna
TPC	Transmission Power Control
ULP-AMI	Ultra Low Power Medical Implant
ULP-AMI-P	Ultra Low Power Medical Implant Peripherals
UWB	Ultra-Wideband
VNA	Vector Network Analyser
WBAN	Wireless Body Area Network
WLAN	Wireless Local Area Network
WMTS	Wireless Medical Telemetry System
WPAN	Wireless Personal Area Network
WSN	Wireless Sensor Network
VSWR	Voltage Standing Wave Ratio

Contents

1

Background

1.1 Introduction

Body area network (BAN) was first discussed under the umbrella of personal area network (PAN). Zimmermann is credited with the invention of the concept of BAN based on his work at Massachusetts Institute of Technology (MIT) and later at International Business Machines (IBM) [1]. He discussed a combination of portable computing devices and short-range wireless link as providing a new paradigm for computing and communication. The link can be established through handshake and communication was made by direct touch by close vicinity (<2 m). In the first version of The Book of Vision of Wireless World Research Forum (WWRF), PAN was shown as the innermost sphere near to the users [2]. In 2004, BAN was described as immediate environment around people which includes those "nearest" object that might be part of body [9].

Task Group 15.6 (TG6) of the IEEE 802 Local and Metropolitan Area Network Standards Committee [3] is the first technical group that announced to make an international standard on wireless BAN (WBAN) [4]. From an evolutional point of view on wireless communication networks within IEEE 802 Local and Metropolitan Area Network Standards committee, WBAN is a natural extension of a chain from wireless metropolitan area network (WMAN), wireless local area network (WLAN), and wireless personal area network (WPAN).

A conceptual sketch in the sense of communication ranges of the above wireless networks is shown in Figure 1.1. WBAN provides a unique solution

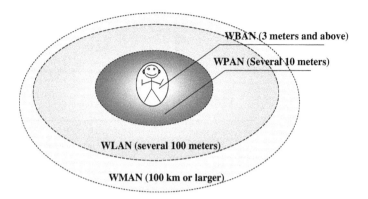

Fig. 1.1 Communication ranges of WMAN, WLAN, WPAN, and WBAN.

of short-range communications to connect devices distributed in, on, and in the peripheral proximity around human body with consideration for human body safety. It targets a convergence of sensors/actuators and wireless communication in medical, healthcare, and body-centric electronic devices

In the above wireless networks, WBAN and WPAN are close in concepts and technologies. However, there are a number of differences between them. A WBAN device can be physically on the surface or inside of a person's body, the safety to body organs and tissues is of most important concern in comparison to all other wireless networks. Moreover, WBAN channel presents different characteristics compared to that of WPAN. The body surface channels mainly depend on space wave and surface wave propagation [5]. Furthermore, the implant communication is not considered in WPAN. As tissue medium of humans is lossy and mainly consists of salt water, the propagation of electromagnetic wave attenuates much faster than that in free space [6]. Since WBAN devices are physically on the surface or inside of a body, the antenna pattern may be affected by the body.

Recent technological advances in low-power microelectronics, miniaturization, and wireless networking enable the design and proliferation of WBAN. However, engineers and designers of WBAN may face a number of challenging tasks such as regulatory circumstance, channel model, low power consumption, form factor, thermal effect, antenna and body loss, high-efficiency radios, reasonable data rate, high reliability, efficient medium access control, etc. This book is going to address some of these main challenges.

1.2 WBAN and Medical/Healthcare Monitoring

One of the most promising applications of WBAN is for human health monitoring and medical vigilance. A number of tiny wireless sensors, strategically
placed on/in the human body, create a WBAN that can monitor various vital
signs, providing real-time feedback to the user and medical personnel. The
WBAN promises to revolutionize health monitoring medical vigilance.

Recent technological advances in wireless networking, microelectronics
integration and miniaturization, sensors, and the Internet allow us to funda-
mentally modernize and change the way medical and healthcare services are
deployed and delivered. Focus on prevention and early detection of disease
or optimal maintenance of chronic conditions promise to augment existing
healthcare systems that are mostly structured and optimized for reacting to
crisis and managing illness rather than wellness.

Wearable systems for continuous biosignal monitoring are a key technology in helping the transition to more proactive and affordable healthcare and
medical vigilance. They allow an individual to closely monitor changes in
her or his vital signs and provide feedback to help maintain an optimal health
status. These systems can even alert medical personnel when life-threatening
changes occur [7, 8].

During the last few years there has been a significant increase in the number
and variety of wearable health-monitoring devices, ranging from simple pulse
monitors, activity monitors, and portable Holter monitors, to sophisticated
and expensive implantable sensors. However, wider acceptance of the existing
systems is still limited by the following important restrictions.

Traditionally, personal medical monitoring systems, such as Holter monitors, have been used only to collect data. Data processing and analysis are
performed offline, making such devices impractical for continual monitoring
and early detection of medical disorders. Systems with multiple sensors for
physical rehabilitation often feature unwieldy wires between the sensors and
the monitoring system. These wires may limit the patient's activity and level
of comfort and thus negatively influence the measured results. One of the most
promising approaches in building wireless health-monitoring systems makes
use of WBAN.

Advanced medical and healthcare applications can be extended to medical
closed loop control. Figure 1.2 shows an intuitive view of automatic medical

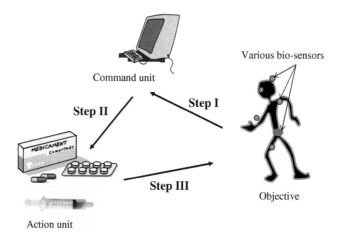

Fig. 1.2 Illustration of closed loop control.

treatment/dosing process with closed loop control. At the first step, various vital and healthcare data are collected using different sensors attached to a person.

The collected data is sent to a command unit. At the second step, the command unit decides the corresponding treatment method or appropriate dosing based on the received vital and healthcare data. Then, the command unit sends a command to the action unit. At the third step, the action unit applies the treatment or dosing to the objective. When the treatment or dosing is finished, sensors will collect updated vital and healthcare data and the process enters another circulation.

The medical motivation is to save life and to improve the health outcomes with easy, fast diagnosis, and treatment. WBAN can be used as an efficient tool in assisting efficient medical activities. The goal for homecare services is to improve quality of life and independence for patients by supporting care at home. The WBAN environment at home is to replace expensive hospital-based care with homecare patient support.

1.3   Other Applications

There are plenty of applications and usage models that can be facilitated by using WBAN. Except for medical and healthcare monitoring mentioned in

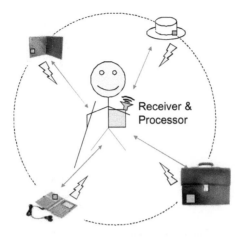

(a) Belongings check

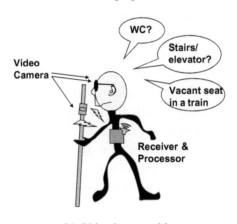

(b) Objective searching

Fig. 1.3 Assistance to people with visual disability.

last section, assistance to people with disabilities and body-centric consumer electronics are other two attractive fields for WBAN.

Figure 1.3 describes applications for assisting people with visual disability. In Figure 1.3(a), WBAN sensors are attached to the belongings of a person with visual disability. A reasonable threshold of range between these sensors with a receiver carried by the person is set in advance. If the person forgets to take corresponding belongings and leaves them at a distance that exceeds the pre-set threshold, the receiver will send warning signal to remind the person.

Fig. 1.4 Audio/Video source sharing with WBAN.

Furthermore, in advanced applications, cameras can be attached to a people with visual disability, either on glasses and/or on the stick. Video pictures taken by cameras are sent to a receiver carried by the people, where visual pictures are converted to voice to provide guidance to the people as shown in Figure 1.3(b). The similar principle can be used for assisting people with speech disability. In this case, sensors to catch finger and hand movements are used. The obtained information caught by sensors is converted into speech.

Examples of body-centric electronics include user interface, wireless headphone, audio or video streaming, game controller with action sensors, entertainment data, etc. Using WBAN in these applications can not only increase convenience by deleting wires but also provide a method of source sharing. For example, two or more users can share a same music player or video source as shown in Figure 1.4.

1.4   Overview of Book

The goal of this book is to provide with a basis for innovative design of WBAN with insights into main aspects of WBAN. The book covers various topics such as frequency regulations, antenna, radio wave propagation and human tissues, modulations and channel coding, medium access control (MAC), and standardization review.

Chapter 2 concentrates on frequency regulations. There are a number of frequency bands such as ultra-wideband (UWB), frequency band for industrial, scientific and medical (ISM), medical implant communication service (MICS), wireless medical telemetry system (WMTS). They are considered as candidates for WBAN. Main regulations on corresponding frequency bands in different countries and regions are investigated and summarized.

Chapter 3 provides detail on antenna, propagation, and channel modeling. The chapter deals with UWB antenna design for WBAN and its differences with free space design one. Effects of radio frequency on tissues and organs and effects of human tissues on RF propagations will be addressed. Moreover, brief discussions on numerical methods are provided. The chapter also provides detail information on specific absorption rate (SAR) and thermal effects. Channel modeling in three different environments, anechoic camber, office room, and hospital room, have been discussed and path loss model and power delay profile are provided.

Chapter 4 deals with physical (PHY) layer technologies. Basic concept of WBAN from the PHY point of view is illustrated. Brief description on general transceiver structure of digital communication systems and calculation of wireless system link budget are provided. Both narrow band and UWB modulations are discussed with the emphasis being laid on those suitable for WBAN. Simple convolutional codes and block codes easily to be implemented are described. Several typical short-range radios are compared for their suitability for WBAN.

Chapter 5 presents medium access control (MAC) technologies for WBAN. The chapter deals with the requirements and new challenges of WBAN for medical applications. The coexistence of Institute of Electrical and Electronic Engineers (IEEE) wireless local area networks (WLAN) and wireless personal area networks (WPAN) has been discussed. The chapter also provides channel fading analysis from MAC point of view. A unified MAC design which is independent of underlying PHY technologies is given.

Chapter 6 gives brief overviews on the activities of several ongoing standardizations such as IEEE 802.15.6, IEEE11073, and eHealth project of the European Telecommunication Standards Institute (ETSI).

References

[1] T. G. Zimmerman, "Personal area network: Near-field intrabody communication," *IBM Systems Journal*, vol. 38, no. 4, pp. 566–574, 1999.
[2] Wireless World Research Forum, The Book of Vision 2001: Vision of the Wireless World 2001.
[3] http://grouper.ieee.org/groups/802/15/.

[4] A. W. Astrin, H.-B. Li, and R. Kohno, "Standardization for body area networks," *IEICE Transactions on Communications*, vol. E92-B, no. 2, pp. 366–372. Feb. 2009.

[5] A. Fort, C. Desset and P. De Doncker, *et al.*, "An ultra-wideband body area propagation channel model — from statistics to implementation," *IEEE Transactions on Microwave Theory and Techniques*, vol. 54, no. 4, pp. 1820–1826, 2006.

[6] S.K.S. Gupta, S. Lalwani, and Y. Prakash, *et al.*, "Towards a propagation model for wireless biomedical applications," *IEEE International Conference on Communications*, vol. 3, pp. 1993–1997, 2003.

[7] Lorincz, K., D. Malan, T. R. F. Fulford-Jones, A. Nawoj, A. Clavel, V. Shnayder, G. Mainland, S. Moulton, and M. Welsh, "Sensor networks for emergency response: Challenges and opportunities," *IEEE Pervasive Computing, Special Issue on Pervasive Computing for First Response*, vol. 3, pp. 16–23, October–December 2004.

[8] Martin, T., E. Jovanov, and D. Raskovic, "Issues in wearable computing for medical monitoring in applications: A case study of a wearable ECG monitoring device," *Proceedings of ISWC* 2000.

[9] P. Coronel, W. Schott, K. Schwieger, E. Zimmermann, H. Maass et al., "Wireless Body Area and Sensor Networks," *WWRF Briefings*, 2004.

2

Regulations

Like other wireless systems, wireless body area network (WBAN) operates on radio wave and is subjected to frequency spectrum regulations. The global organization for frequency spectrum management is the International Telecommunication Union Radio Communication Sector (ITU-R) [1], while there are local administrative organizations in countries or regions such as Federal Communications Commission (FCC) of United States of America [2], European Commission (EC) of European Union [3], Ministry of Industrial and Information Technology (MIIT) of China [4], Ministry of Internal Affairs and Communications (MIC) of Japan [5], etc.

Besides the frequency spectrum regulations, radiation protection safety levels are of strong concerns because that WBAN is operated in the proximity of, on, or inside the human body. International Commission on Non-ionizing Radiation Protection (ICNIRP) [6] provides sound guidance on the health hazards of non-ionizing radiation exposure. A popular related terminology is the specific absorption rate (SAR), which describes the potential for heating of the body tissues when exposed to a radio frequency (RF) electromagnetic field. SAR is defined as the RF power absorbed per unit of mass of tissue, and is measured in watts per kilogram (W/kg). International Electrotechnical Commission (IEC) [7] issued technical guidelines on SAR, while different countries or regions have their own regulations in accordance with IEC. Moreover, IEEE Std 11073-00101™-2008 delivered guidelines for medical devices using RF communications [8]. The issues of SAR will be covered in Chapter 3. In this chapter, the emphasis will be put on frequency spectrum regulations.

Because that WBAN operates in a limited space around a body, wireless technologies that provide short-range communication link with low radiation level and low consumption power are favored. Good candidate technologies include ultra-wideband (UWB) radios, wireless personal area network (WPAN) on industrial, scientific, and medical (ISM) band, short-range radio devices, etc. Already, assigned frequency spectrums for medical purpose of WBAN are available, such as frequency band for medical implant communications service (MICS) and frequency band for wireless medical telemetry system (WMTS).

2.1   Regulations on UWB

UWB uses very low power emission levels with a large portion of the radio spectrum to provide either short-range but high data rate communications, or low data rate communications but high precision ranging. The first UWB regulation was issued by FCC in February 2002 [9] as an outcome of deliberation in connection with the National Telecommunications and Information Administration (NTIA). This UWB regulation opened the door for UWB commercial products without licenses.

Because of the potential of broad application areas and huge markets, other regulation authorities started UWB regulation procedure shortly after FCC and UWB regulations have been announced in a number of countries and regions. However, regulations are different from each other for reasons of local spectrum management. Although UWB regulations include applications such as radar and imaging, the focus of this book is UWB regulation on communications. Hence, only UWB regulations related to communication purpose are overviewed. It should be noted that some regulators, such as EU, Japan, and Korea, are in the procedure to update their regulation on UWB, while some countries, such as Australia and New Zealand, have not completed UWB regulation.

2.1.1   FCC's Regulation on UWB

UWB is defined in terms of radiated signal bandwidth. The radiated signal from a UWB transmitter must occupy a bandwidth that exceeds the lesser of 500 MHz or 20% of fractional bandwidth. The fractional bandwidth (FBW) is defined as:

$$\text{FBW} = 2\frac{F_H - F_L}{F_H + F_L}, \tag{2.1}$$

where F_H and F_L are the upper and lower boundaries of the UWB signal spectrum, respectively. Both are defined at the points that are 10 dB below the highest radiated emission level. In the frequency range of 3.1–10.6 GHz, which is considered suitable for communication purposes, FBW is always larger than 500 MHz. Therefore, 500 MHz can be used as the measure of minimum bandwidth for UWB.

The equivalent isotropic radiated power (EIRP) for a UWB transmitter is restricted by a power spectrum density (PSD) mask. For communication purposes, UWB can only operate either indoor or on handheld. Mean PSD masks for both indoor and handheld are shown in Figure 2.1. It can be seen that PSD masks are the same in the frequency range of 3.1–10.6 GHz. However, the PSD mask for handheld is 10 dB lower than that for indoor in the range of 1.61–3.1 GHz and beyond 10.6 GHz.

Another specific technical requirement for handheld UWB system is that a handheld device shall cease transmission within 10 seconds unless it receives a confirming acknowledgment from the associated receiver that its transmission is being received.

Besides the PSD mask shown in Figure 2.1, there is simultaneously a limit on peak power emission. The highest radiated EIRP within a 50-MHz bandwidth must be smaller than or equal to 0 dBm/MHz. This peak power limit is also adopted in other countries' regulations. When measuring the peak

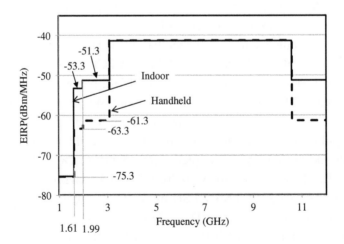

Fig. 2.1 FCC's mean PSD mask for UWB.

power, the resolution bandwidth used shall not be lower than 1 MHz. The peak EIRP limit shall be calculated using the following formula:

$$\text{EIRP}_{[\text{BW}=50 \text{ MHz}]} = 20 \, \log \left(\frac{\text{RBW}}{50} \right), \tag{2.2}$$

where RBW is the resolution bandwidth in megahertz used in the measurement.

FCC's regulation on UWB is summarized in Subpart F of Part 15 of 47 Code of Federal Regulations (CFR). One thing needs to be keep in mind is that all UWB devices may not be employed for the operation of toys. Operation onboard an aircraft, a ship, or a satellite is prohibited.

2.1.2   UWB Regulation in Europe

There are three regulation/standardization bodies which are responsible in forming European regulation and standards. They are European Union (EU), Conference of European Post & Telecommunication (CEPT), and European Telecom Standards Institute (ETSI). The most recent updated documents issued from these three organizations are, respectively, Decision 2009/343/EC [10], CEPT Report 34 [11], and Draft ETSI EN 302 065 V1.2.1 [12]. Main regulations and technical requirements for generic UWB usage are described in the following. It should be noticed that hereafter we only describe the mean EIRP mask, because UWB regulations in EU and other countries adopt the same peak power emission limit as that of FCC.

EU gives a different definition on UWB bandwidth compared to that of FCC. The minimum operating bandwidth is defined as 50 MHz, which is much smaller than the 500 MHz bandwidth defined by FCC. However, same as FCC, bandwidth is measured at the points that are 10 dB below the highest radiated emission level. The mean PSD mask for indoor UWB defined in EU is shown in Figure 2.2. At UWB low band, the maximum EIRP of −41.3 dBm/MHz can be applied in the frequency range of 3.1–4.8 GHz if low duty cycle (LDC) or detect-and-avoid (DAA) is implemented. Without LDC or DAA, an EIRP of −70 dBm/MHz (−80 dBm/MHz from 3.4 to 3.8 GHz) must be satisfied. However, the −41.3 dBm/MHz PSD mask can be applied in the frequency range of 4.2–4.8 GHz without LDC or DAA until the end of 2010. At UWB high band, the maximum EIRP is −41.3 dBm/MHz in the range of 6–8.5 GHz. In the range of 8.5–9 GHz, the same value can be applied with DAA or

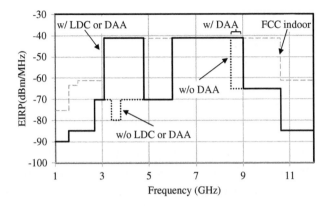

Fig. 2.2 Mean PSD mask in EU' UWB regulation.

Table 2.1. Maximum EIRPs and thresholds in zone definitions.

Victim system and frequency band (GHz)	Max. EIRP in Zone 1 (dBm/MHz)	Threshold 1 (dBm)	Max. EIRP in Zone 2 (dBm/MHz)	Threshold 2 (dBm)	Max. EIRP in Zone 3 (dBm/MHz)
Radiolocation S-band: 3.1–3.4	−70	−38	−41.3	—	—
BWA 3.4–3.8	−80	−38	−65	−61	−41.3
Radiolocation X-band: 8.5–9	−65	−61	−41.3	—	—

the maximum EIRP of −65 dBm/MHz must be satisfied without DAA. Both DAA and LDC are interference mitigation techniques to reduce interference from UWB devices to other existing wireless systems.

A flexible DAA mechanism is introduced, in which UWB is allowed with different emission levels according to their operating zones. The operating zones are divided by agreed thresholds of detected victim signals. Zone threshold levels and permitted EIRP emission levels in each zone are decided for three victim systems; an S-band radiolocation system at 3.1–3.4 GHz, a broadband wireless access (BWA) system at 3.4–3.8 GHz, and an X-band radiolocation system at 8.5–9 GHz. Table 2.1 shows the corresponding values of thresholds and EIRPs for each system. There are only two zones for each of the two radiolocation systems, while there are three zones for the BWA system.

Duty cycle defines the percentage or transmission time of a UWB device in a given time period. LDC parameters are first defined in EU. The LDC parameters are as follows:

- Maximum Tx on $\leq$ 5 ms.
- Minimum mean Tx off $\geq$ 38 ms (mean value averaged over 1 second).
- Accumulated minimum Tx off (Σ Tx off) $\geq$ 950 ms in 1 second.
- Maximum accumulated transmission time (Σ Tx on) 18 s in 1 hour.

Transmission power control (TPC) is an interference mitigation technique required by EU for UWB radio devices intended to be installed in road or rail vehicles. UWB devices with TPC function can apply the maximum EIRP of -41.3 dBm/MHz in frequency range of 3.1–4.8 GHz, 6–8.5 GHz, and 8.5–9 GHz. Otherwise, the maximum EIRP of -53.3 dBm/MHz must be satisfied. We summarize the required interference techniques to apply the maximum EIRP of -41.3 dBm/MHz in Table 2.2.

Table 2.2.   Required interference techniques to apply the -41.3 dBm/MHz mask.

	Frequency bands (GHz)		
	3.1–4.8	6–8.5	8.5–9
Indoor usage	LDC or DAA	No need	DAA
In road or rail vehicles	LDC or DAA+TPC	LDC or TPC	DAA+LDC or DAA+TPC

2.1.3   UWB Regulation in Japan

UWB regulation was enforced in August 2006 as Article 4.4 of Regulations for Enforcement of Radio Law issued by MIC Japan. We summarize the main parts of the general technical requirements in the following. Because the peak power limit is the same as USA and EU, we only describe the mean PSD mask and which is shown in Figure 2.3 with FCC's PSD mask being given as a comparison.

The available frequency ranges are 3.4–4.8 GHz in low band and 7.25–10.25 GHz in high band. In low band, interference mitigation techniques are required although no descriptions on the specifications of interference mitigation techniques are given. Interference mitigation techniques are not required in the high band. Moreover, the frequency band of 4.2–4.8 GHz can be used without interference mitigation techniques until December 2010. However, the Japanese regulators are considering to extend the above deadline from

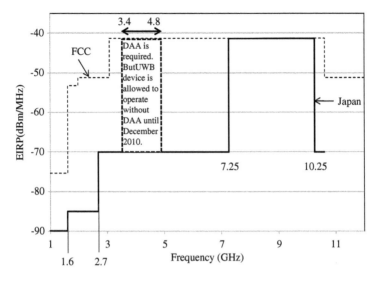

Fig. 2.3 Mean PSD mask in Japanese UWB regulation.

December 2010 to December 2013. The following are some specific technical requirements in Japanese UWB regulation.

- As an assurance of indoor operation, a UWB radio equipment can start transmission at any time if it is connected to the AC mains power supply. A UWB radio equipment not connected to the AC mains power supply shall be permitted to emit radio waves only after it receives a signal from another radio equipment connected to the AC mains power supply.
- Transmission data rate shall be higher than 50 Mbps excepting for such cases as noise or interference from other radio stations needs to be avoided.
- Frequency bandwidth between the upper and lower frequencies for which the radiation power drops 10 dB below the maximum radiation power shall be 450 MHz or more.

Discussion on specifications of necessary interference mitigation techniques and other regulation parameters is still on going. It is expected that some modifications or updates may be introduced in Japanese UWB regulation around the end of 2010.

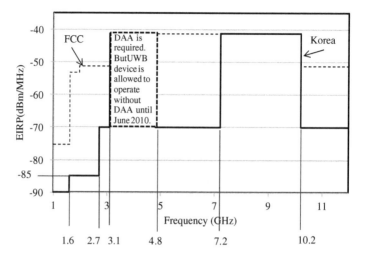

Fig. 2.4 Mean PSD mask in Korean UWB regulation.

2.1.4   UWB Regulation in Korea

Korean UWB regulation came into force in the summer of 2006. The PSD mask is similar to that of Japan, while the frequency ranges are a little different. Some parameters like LDC of the interference mitigation techniques are similar to that of EU. The mean PSD mask in Korea is shown in Figure 2.4.

The available frequency ranges are 3.1–4.8 GHz in low band and 7.2–10.2 GHz in high band. Interference mitigation techniques are required in the low band, while there is no need of interference mitigation techniques in the high band. However, the Korean regulators are in discussion to waive the interference mitigation technique requirement to December 2013 or even beyond. Main specific regulations parameters and technical requirements are as follows:

- UWB equipment is prohibited to operate on aeroplanes, ships, satellites, and radio-controlled flying models.
- Frequency bandwidth between the upper and lower frequencies for which the radiation power drops 10 dB below the maximum radiation power shall be 450 MHz or more.
- One of the following interference mitigation techniques can used in the frequency band of 3.1–4.8 GHz.

(1) The mean PSD shall not exceed −70 dBm/MHz.

(2) As effective LDC parameters, continuous transmission interval shall not exceed 5 ms and stop transmission interval shall be larger than 1 second.

(3) The emission power of a UWB device shall be reduced to a level smaller than −70 dBm/MHz within 2 seconds when it detected a victim signal level larger than −80 dBm/MHz.

(4) A UWB device shall avoid within 2 seconds when it detected a victim signal level larger than −80 dBm/MHz.

2.1.5 UWB Regulation in China

UWB regulation was enforced in December 2008 by MIIT in China. The UWB signal bandwidth, between the upper and lower frequencies for which the radiation power drops 10 dB below the maximum radiation power, must exceed 500 MHz. The available frequency range to apply a −41 dBm/MHz PSD mask is 6–9 GHz. Another frequency range can apply the −41 dBm/MHz PSD mask is 4.2–4.8 GHz with proper interference mitigation techniques. The interference mitigation techniques are waived until December 2010. The specifications and technique requirements of the interference mitigation techniques are still under discussion.

The mean PSD mask in China is shown in Figure 2.5. Other main regulation parameters and technical requirements are as follows:

- UWB equipment is prohibited to operate on aircrafts.
- UWB equipment is prohibited to operate in around 1 km range of the radio-astronomy stations.
- The management for UWB equipment is following the micropower (short-range) radio Tx. equipment radio regulatory management requirements in China. UWB equipment shall obtain the type of approval certification for the Tx equipment by MIIT before putting into service.
- UWB Tx equipment shall not create any interference to other radio service stations. UWB shall not ask any interference protection from other radio service stations.

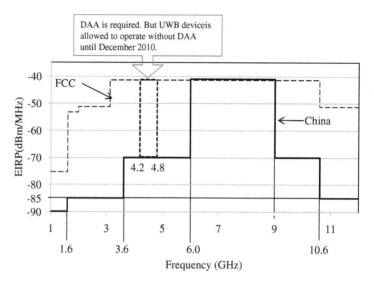

Fig. 2.5 Mean PSD mask in Chinese UWB regulation.

2.2 Regulations on ISM Bands

2.2.1 Frequency Bands Allocated to ISM

ISM bands are internationally reserved frequency spectrums to be used for ISM purposes. Because radio devices using ISM bands are generally allowed for unlicensed operation, ISM band can also be regarded as one of the frequency candidates for WBAN.

International frequency allocations for ISM bands by ITU are contained in the radio regulation (RR) Nos. 5.138, 5.150, and 5.280. The main results are summarized in Table 2.3. Individual countries' use of these bands designated in these sections may differ due to variations in national radio regulations.

In Table 2.3, Region 1 and Region 2 are following ITU definition. Region 1 comprises Europe, Africa, the Middle East, west of the Persian Gulf including Iraq, the former Soviet Union, and Mongolia, while Region 2 covers the Americas, Greenland, and some of the eastern Pacific Islands. It should be noted that 433.05–434.79 MHz are assigned in Region 1 except in the countries mentioned in No. 5.280.

Table 2.3. Designated frequency bands and corresponding sections in ITU-RR.

Frequency band	Center frequency	Sections in ITU RR
6.765–6.795 MHz	6.78 MHz	5.138
13.553–13.567 MHz	13.567 MHz	5.150
26.957–27.283 MHz	27.12 MHz	5.150
40.66–40.70 MHz	40.68 MHz	5.150
433.05–434.79 MHz	433.92 MHz	5.138 (Region 1), 5.280
902–928 MHz[1]	915 MHz	5.150 (Region 2)
2400–2500 MHz	2450 MHz	5.150
5.725–5.825 GHz	5.8 GHz	5.150
24.00–24.25 GHz	24.125 GHz	5.150
61.00–61.50 GHz	61.25 GHz	5.138
122–123 GHz	122.5 GHz	5.138
244–246 GHz	245 GHz	5.138

[1] 896 MHz in the United Kingdom.

Table 2.4. Technical Characteristics in ECC/DEC/(04)02.

Frequency band (MHz)	Emission power	Channel spacing	Duty cycle[1]
433.050–434.790	10 mW ERP	No channel spacing	Below 10%
433.050–434.790	1 mW ERP −13 dBm/10 kHz for wideband channels[2]	No channel spacing	No restriction
434.040–434.790	10 mW ERP	Up to 25 kHz	No restriction

[1]The ratio of the maximum transmitter "on" time on one carrier frequency in a one-hour period.
[2]A bandwidth greater than 250 kHz.

2.2.2 ISM Frequency Bands Suitable for WBAN

Frequency bands that are most suited for WBAN at ISM bands include 433.05–434.79 MHz, 902–928 MHz, and 2400–2500 MHz.

2.2.2.1 433.05–434.79 MHz

This frequency band is available in EU, Australia, and some other countries, but with different restrictions. In Australia, the available band is 434.05–434.79 MHz and restriction is only given on maximum EIRP which is 25 mW. In EU, the regulation is summarized in ECC/DEC/(04)02 for non-specific short-range devices with the technical characteristics as shown in Table 2.4, in which the emission power is defined as effective radiated power (ERP).

Table 2.5.   Australian regulation on 915–928 MHz.

Transmitter types	Maximum EIRP	Limitations
All transmitters	3 mW	
Digital modulation transmitters	1 W	1. The radiated peak power spectral density in any 3 kHz is limited to 25 mW per 3 kHz. 2. The minimum 6 dB bandwidth must be at least 500 kHz.
Frequency hopping transmitters	1 W	A minimum of 20 hopping frequencies must be used.

2.2.2.2   902–928 MHz

This band can be used without license in North America, Australia, New Zealand, and a number of other nations. However, local regulations are different on how to use the spectrum. In USA, unlimited radiated energy within 902–928 MHz is permitted, although there are restrictions on out-band emission levels. The regulation on ISM band of USA is managed by Part 18. In comparison, Australian regulation set restrictions on emission power according to transmitter types, which are summarized in Table 2.5. Moreover, the available band in Australia is 915–928 MHz.

2.2.2.3   2400–2500 MHz

2400–2500 MHz is one of the most popular ISM bands. In the original allocation, this band is assigned for usages of microwave oven, industrial heating equipment, ionization of gases, and so on. Later on, the band is widely used for radio frequency identification (RFID) and radio-controlled flying models. Moreover, a number of existing standards also make use of this band such as, IEEE 802.11b/g for wireless local area network (WLAN), IEEE 802.15.1 which defines the specifications of physical layer for Bluetooth [13], IEEE 802.15.4 which defines the specifications of physical and media access control (MAC) layers for ZigBee [14], and the physical layer of chirp spread spectrum (CSS) defined in IEEE 802.15.4a.

The use of ISM band is subject to authorization of local administrations. Co-existence issue must be paid enough attentions, because the radio communication services operated at this band basically are not protected from interference. Moreover, besides the frequency regulation on ISM band, there are also local regulations for unlicensed communication devices. Those rules for unlicensed communication devices must be taken into consideration.

2.3 Frequency Bands for Medical Applications

2.3.1 Medical Implant Communications Service

Medical implant communications service (MICS) uses a radio communication system, with which data in support of diagnostic or therapeutic functions associated with medical implemented devices (MID) are transmitted via RF communications between MID and a remote monitoring unit.

Due to the requirement of a single available band worldwide for MICS, ITU-R recommended the operation of MICS on a secondary basis in the band of 401–406 MHz [15]. The EIRP is limited to -16 dBm (25 μW) in a reference bandwidth of 300 kHz. In order to avoid interferers and support the simultaneous operation of multiple devices in the same area (such as clinics with multiple rooms), ITU-R recommended a total required bandwidth of 3 MHz to support 10 channels. It was also pointed out that only one or two channels will be usable in many environments.

Local administrative organizations made different regulations on how to use the MICS band recommended by ITU-R. Generally, the MICS band is permitted on a license-by-rule basis.

2.3.1.1 MICS band in USA

Originally, FCC only assigned 402–405 MHz for MICS as a rule, which took effect on January 14, 2000. After debates with representatives of the medical communities and equipment manufacturers, it was approved that the establishment of MICS would greatly improve the utility of medical implant devices by allowing physicians to establish high-speed, easy-to-use, reliable, short-range (2 m) wireless links to connect such devices with monitoring and control equipment. Based on the above observation, FCC intended to expand the MICS band and called the expanded band as MedRadio band. FCC released notice of proposed rulemaking (NPRM) and notice of inquiry (NOI) [16] on July 18, 2006. The rules changing the name of the service to MedRadio and expanding the designated frequency band were contained in 47 CFR 95.628 and took effect on August 12, 2009.

Table 2.6. FCC's regulation of MedRadio on MICS.

Frequency band (MHz)	Bandwidth limits (kHz)	Emission power limits (EIRP)	Remarks
401–401.85	100	250 nW	(1) Non-LBT
401.85–402	150	25 μW	(2) 0.1% duty cycle and maximum 100 communication sessions per hour
402–405	300	25 μW	With LBT
403.5–403.8	300	100 nW	(1) Non-LBT
			(2) 0.01% duty cycle and maximum 10 communication sessions per hour
405–406	100	250 nW	(1) Non-LBT
			(2) 0.1%duty cycle and maximum 100 communication sessions per hour

Within the band for MedRadio for MICS, a channel access method of monitoring the usage of frequency before use it, also known as, listen-before-talk (LBT) is required. There are exceptions for LBT with certain conditions. Details are summarized in Table 2.6.

The non-LBT operation conditions harmonize with the EU regulation, that is described in the next sub-section.

2.3.1.2   MICS band in EU

EU's decision on MICS was issued in ERC/DEC/(01)17 on March 12, 2001 [17]. Instead of MICS, it is referred to as ultralow power medical implant (ULP-AMI) and peripherals (ULP-AMI-P). Then, the operating conditions were summarized in two standards, EN 301 839 [18] and EN 302 537 [19], issued by European Telecom Standards Institute (ETSI), respectively, in January 2009 and December 2007. EN 301 839 provides the guideline on 402–405 MHz, while EN 302 537 provides guideline on 401–402 MHz and 405–406 MHz.

The operation conditions of ULP-AMI and ULP-AMI-P are summarized in Table 2.7. Besides the conditions in the table, it is also required that the length of a transmission on a channel as a result of a single medical implant event (MIE) is limited to 30 seconds per hour if LBT is not applied. LBT can be done either by a ULP-AMI device or a ULP-AMI-P device.

Table 2.7. Operating conditions of ULP-AMI and HLP-AMI-P.

Frequency band (MHz)	Bandwidth limits (kHz)	Emission power limits (ERP)	Remarks
401–402	100	250 nW	0.1% duty cycle and maximum 100 repetitive transmission per hour
401–402	100	25 μW	With LBT
402–405	300	25 μW	With LBT
403.5–403.8	300	100 nW	(1) Non-LBT (2) 0.01% duty cycle and maximum 10 communication sessions per hour
405–406	100	250 nW	0.1% duty cycle and maximum 100 repetitive transmission per hour
405–406	100	25 μW	With LBT

2.3.1.3 MICS band in Australia, Japan, Korea, and others

MICS regulations have also been being conducted in other countries or regions. We simply summarize the regulation status of Australia, Japan, and Korea in the following.

MICS regulations of Australia, Japan, and Korea are similar to that of EU. However, only 402–405 MHz is permitted and there are also differences in available bands and operating conditions. Details are summarized in Tables 2.8–2.10.

It should be noticed that Australia, Japan, and Korea define 403.5–403.8 MHz band (403.56–403.76 MHz for Australia) for medical implant telemetry system (MITS). In Japan, this band is only allowed for unidirectional communication, i.e., from an implant device to an outside body device. In Australia, MICS band is allocated within the category of low interference potential devices (LIPD). In Japan, MICS is allocated within the category of specific low power equipments. Generally, the emission power of specific low power equipments is below 0.01 W as defined in Japanese regulation. Recently, modification of the maximum emission power from 0.01 W to 1 W is under evaluation.

The Asia-Pacific Telecommunity Wireless Forum (AWF) made a recommendation to Asia-Pacific Telecommunity (APT) to allow the use of the band 402–405 MHz for MICS and 403.5–403.8 MHz for MITS [20], in which,

Table 2.8.   MICS regulation of Australia.

Frequency band (MHz)	Bandwidth limits (kHz)	Maximum EIRP	Remarks
402–405	300	25 μW	With LBT, systems must have a minimum of nine channels selectable
403.56–403.76	200	100 nW	Non-LBT

Table 2.9.   MICS Regulation of Japan.

Frequency band (MHz)	Bandwidth limits (kHz)	Maximum EIRP	Remarks
402–405	300	25 μW	With LBT,
403.5–403.8	300	100 nW	(1) Non-LBT (2) 0.01% duty cycle and maximum 10 communication sessions per hour (3) Unidirectional transmission

Table 2.10.   MICS regulation of Koreas.

Frequency band (MHz)	Bandwidth limits (kHz)	Maximum EIRP	Remarks
402–405	300	25 μW	With LBT, systems must have a minimum of nine channels selectable
403.5–403.8	300	100 nW	Non-LBT

407–425 MHz is assigned as additional bands for MICS, because this band is chosen by some administrations.

2.3.2   Wireless Medical Telemetry System

Wireless medical telemetry system (WMTS) is a radio communication system, with which the measured physiological parameters and other patient-related information are transmitted via RF communication between a patient-worn transmitter and a remote monitoring unit. By applying WMTS, it would greatly increase the convenience for both patients and medical or healthcare workers.

FCC of USA and MIC of Japan assign specific frequency spectrums for WMTS. Australia and New Zealand define a so-called biomedical telemetry transmitter and assign specific spectrums accordingly.

2.3.2.1 WMTS bands in USA

The FCC rules and regulations are codified in Subpart H of Title 47 of the CFR. Three frequency bands are specified to WMTS operations, i.e., 608–614 MHz, 1395–1400 MHz, and 1427–1429.5 MHz. However, in some locations 1427–1429.5 MHz cannot be used. Instead, 1429–1431.5 MHz can be used for WMTS in those locations. The interested readers can find the information in CFR Part § 90.259(b) (4). Generally, WMTS should only operate within hospitals.

Main operating conditions of WMTS as regulated by FCC are summarized in Table 2.11.

Table 2.11. WMTS regulation of FCC.

Frequency band (MHz)	Channel allocation (MHz)	Field strength limits	Remarks
608–614	608.0–609.5 609.5–611.0 611.0–612.5 612.5–614.0[1]	200 mv/m	Measured at a distance of 3 m with a quasi-peak detector
1395–1400 1427–1429.5	Channel is not specified	740 mv/m	Measured at a distance of 3 m with an averaging detector and a 1 MHz measurement bandwidth

[1] It is allowed to combine the four 1.5 MHz channels at 608–614 MHz to generate a single 6 MHz channel.

It should be noticed that FCC's regulation on MedRadio also allows WMTS operation. Referred to as body-worn devices, they may operate only at 401–402 or 405–406 MHz. The operating conditions described in Table 2.6 are applied. Temporary body-worn devices that are used externally to evaluate the efficacy of a more permanent medical implant device may also operate at 402–405 MHz. Temporary body-worn devices must comply with all rules for MICS. In addition, the maximum EIRP should not exceed 200 nW and operation must be subject to certain time limit.

2.3.2.2 WMTS bands in Japan

In Japan, WMTS is allocated within the category of specific low power equipments. WMTS band is assigned for biosignal telemetry transmission in hospitals, clinics, other medical organizations, and research institutes. It should be noticed that only unidirectional communications, i.e., from a sensor radio device to a monitor radio device, are allowed in Japanese regulation.

However, there are strong demands to enable bidirectional communications for WMTS for the purposes of command or instruction transmission.

WMTS radios are classified into five types according to occupied bandwidth. Permitted emission types are also different. The five types of radios are summarized in Table 2.12.

The frequency bands assigned for WMTS in Japan are 420–430 MHz and 440–450 KHz. Detailed channel allocations for the each of the five types of WMTS radios and available frequency range are summarized, respectively, in Tables 2.13 and 2.14.

Table 2.12.   Classifications of WMTS radios in Japan.

Radio Type	Channel spacing (kHz)	Maximum bandwidth (kHz)	Maximum EIRP	Emission types
A	12.5	8.5	1 mW	FM-FM, 2FSK
B	25	16	1 mW	FM-FM, 2FSK
C	50	32	1 mW	FM-FM, GMSK
D	100	64	1 mW	FM-FM, GMSK
E	500	320	10 mW	FM-FM, PWM[1]-FM, GMSK

[1]PWM: Pulse width modulation.

Table 2.13.   Available channels for WMTS at 420–430 MHz in Japan.

	A	B	C	D	E
Channel centers	420.0500 + $k \times 0.0125$	420.0625 + $k \times 0.025$	420.0750 + $k \times 0.0500$	420.1000 + $k \times 0.1000$	420.3 420.8
Range	420.0500–421.0375	420.0625–421.1125	420.0750–420.9750	420.1000–420.9000	420.3–420.8
Channel centers	424.4875 + $k \times 0.0125$	424.5000 + $k \times 0.025$	424.5125 + $k \times 0.0500$	424.5375 + $k \times 0.1000$	424.7375 425.2375 425.7375
Range	424.4875–425.9750	424.5000–425.9500	424.5125–425.9125	424.5375–425.8375	424.7375 425.7375 425.7375
Channel centers	429.2500 + $k \times 0.0125$	429.2625 + $k \times 0.025$	429.2750 + $k \times 0.0500$	420.3000 + $k \times 0.1000$	429.5
Range	429.2500–429.7373	429.2625–429.7125	429.2750–429.6750	429.3000–429.6000	429.5

$k = 0, 1, 2, \ldots$.

Table 2.14. Available channels for WMTS at 440–450 MHz in Japan.

	A	B	C	D	E
Channel	440.5625	440.5750	440.5875	440.6125	
centers	+	+	+	+	440.8175
	$k \times 0.0125$	$k \times 0.025$	$k \times 0.0500$	$k \times 0.1000$	441.3125
Range	440.5625–441.5500	440.5750–441.5250	440.5875–441.4875	440.6125–441.4125	440.8175
					441.3125
Channel	444.5125	444.5250	444.5375	444.5625	444.7625
centers	+	+	+	+	440.2625
	$k \times 0.0125$	$k \times 0.025$	$k \times 0.0500$	$k \times 0.1000$	
Range	444.5125–445.5000	444.5250–445.4750	444.5375–445.4375	444.5625–445.3625	444.7625
					445.2625
Channel	448.6750	448.6875	448.7000	448.7250	444.7625
centers	+	+	+	+	444.7625
	$k \times 0.0125$	$k \times 0.025$	$k \times 0.0500$	$k \times 0.1000$	
Range	448.6750–449.6625	448.6875–449.6375	448.7000–449.6000	448.7250–449.5250	448.9250
					449.4250

$k = 0, 1, 2, \ldots.$

2.3.2.3 Biomedical telemetry bands in Australia and New Zealand

Two frequency bands are assigned for biomedical telemetry transmitters in Australia [23]. They are (i) 174–230 MHz and (ii) 520–668 MHz. The maximum available emission powers in EIRP for these two bands are, respectively, 10 μW and 3 mW. Because the frequency band (ii) is also used by analogue TV broadcasting, biomedical telemetry transmission is not allowed to be originated from a TV channel.

Biomedical telemetry bands in New Zealand are divided under the category of the general user radio licenses (GURL). There are also two assigned frequency bands of (i) 444–444.925 MHz and (ii) 470–470.5 MHz for biomedical telemetry [24]. It can be seen that the bandwidth for biomedical telemetry is much smaller compared to other regulations. The maximum available emission powers in EIRP for these two bands are, respectively, 25 mW and 100 mW.

2.3.3 Newly Proposed Frequency Band

There are many reasons to apply new frequency bands for medical BAN (MBAN). The current MedRadio bands in USA or ULP-AMI/ ULP-AMI-P in EU have LDC requirements on two wings, i.e., 401–402 MHz and 405–406 MHz. As a result, only the core band of 3 MHz, i.e., 402–405 MHz, is "free" but there is also a restriction on maximum bandwidth of 300 kHz.

Therefore, MedRadio or ULP-AMI bands are not sufficient to support multiple body biosignal sensors in an MBAN, especially for environment like hospitals where many MBANs are simultaneously operated.

In USA, the current WMTS is restricted to and already heavily used in hospitals. Moreover, because of the strong restrictions on body-worn devices operated at 402–405 MHz as described in Section 2.3.2.1, this band is not suitable for body-worn MBAN.

2.3.3.1   Proposed frequency band for MBAN in USA

FCC released NPRM ET Docket No. 08-59 on June 29, 2009 [21]. This NPRM is in response to a petition made by General Electric Healthcare (GEHC) to allow the use of new frequency band for MBAN. The purpose is to enable MBAN for the wireless networking of multiple body sensors used for monitoring a patient's physiological data to increase the quality, flexibility, and efficiency mainly in healthcare facilities.

GEHC's originally proposed frequency band is 2360–2400 MHz with maximum EIRP of 1 mW in 1 MHz bandwidth. The proposed maximum emission bandwidth is 1 MHz.

In addition to GEHC's proposal, three more frequency bands are listed in the NPRM to solicit comments for their suitability for MBAN. These bands are 2300–2305 MHz, 2400–2483.5 MHz, and 5150–5250 MHz. One merit of the GEHC's proposed band is that it is adjacent to the ISM band which may enable low-cost devices. Those technologies developed for ISM band can be applied to the proposed band without difficulty.

2.3.3.2   ETSI's report on new requirements for
frequency band for implants

ETSI released the technical report TR 102 655 V1.1.1 in November, 2008 [22]. This report defines new requirements for radio frequency spectrum usage for low power active medical implant (LP-AMI) and their peripheral radio control systems. The background behind this technical report is that the rapid development and increased use of LP-AMI have triggered new applications that need much higher data rate. Therefore, additional frequency spectrum is required to enable much greater bandwidths for LP-AMI.

Table 2.15. Proposed regulation parameters.

Candidate frequency band (MHz)	Band edge mask width (MHz)	Maximum EIRP (dBm)	LBT	Adaptive power control (APC)	Adaptive frequency selection (AFS)	Minimum number of channels
2483.5–2500	16.5	10	Yes	Yes	Yes	16
2700–3400[1]	20	10	Yes	Yes	Yes	20
2360–2400	20	10	Yes	Yes	Yes	20

[1] 2.7–2.9 GHz may not be not be possible because of allocation of radar.

Three frequency bands are proposed in the technical report in the following order. The proposed regulation parameters for the candidate frequency bands are summarized in Table 2.15.

- From 2483.5 MHz to 2500 MHz.
- A 20-MHz segment within the band of 2700–3400 MHz.
- A 20-MHz segment within the band of 2360–2400 MHz.

It is also pointed out in the report that the following facts need to be taken into consideration. First, both the 2483.5–2500 MHz and 2700–2900 MHz bands are adjacent to the 2.6 GHz mobile band. Second, 2300–2400 MHz band has been assigned for International Mobile Telecommunication (IMT) at ARC-07. Attention must be paid to interference and coexistence issues.

2.4 Short-Range Devices

Because WBAN is a short-range and low-consumption power network, short-range device (SRD) defined in ECC/DEC/(04)02 and ETSI EN 300 220-1 V2.3.1 can also be regarded as candidates for WBAN. Out of the frequencies from 25 MHz to 1 GHz listed in ETSI EN 300 220-1, 400 MHz and 800 MHz frequency bands are of specific interest in the sense of available bandwidths. The 400 MHz bands have been well described in Section 2.2.2.1. We describe the 800-MHz bands for SRD in the following.

The available frequency band at 800 MHz is 863–870 MHz. The general operating rules of SRD at this band are summarized in Table 2.16, while some specified rules are summarized in Table 2.17 [25].

It should also be noticed that the specifications defined in ECC decision or ETSI standard represent the most widely implemented position within the EU and the CEPT countries. However, it should not be assumed that all desig-

Table 2.16. General operating rules of SRD at 800 MHz bands.

Frequency band (MHz)	Maximum ERP	Channel spacing	Modulations	Remarks
863–870[4]	25 mW	≤ 100 kHz[6]	Narrow/wideband modulation	0.1% LDC, or LBT + AFA[2,3,9]
863–870[4]	25 mW Power density is limited to −4,5 dBm/100 kHz[1,7]	—	DSSS and other wideband modulation other than FHSS	0.1% LDC, or LBT + AFA[3,8,9]
863–870[4]	25 mW[1]	≤ 100 kHz[6]	FHSS modulation	0.1% LDC or LBT[2,9]

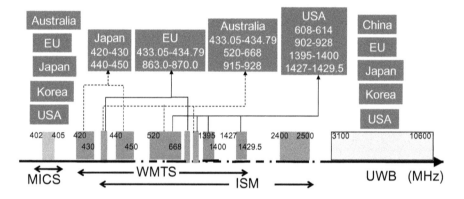

Fig. 2.6 An overview of frequency spectrums for WBAN.

nated bands are available in all countries. Local regulators may have specific requirements or apply specific rules to those bands.

2.5 Summary of Chapter

In this chapter, we summarize the main frequency regulations for WBAN with the strength being put on the available frequency spectrums and operating conditions. Main frequency spectrums investigated are UWB band, ISM band, MICS band, WMTS band, and SRD band. Due to limited resources, it is difficult to include every related regulation. We try to cover representative regulations in this book. All contents are based on published regulations, standards, or documents. Interested authors are suggested to further look into the listed references.

Main available frequency spectrums discussed in this chapter are depicted in Figure 2.6. Because a WBAN operates with people who may travel through

Table 2.17.   Specified operating rules for SRD at 800 MHz bands.

Frequency band (MHz)	Maximum ERP	Channel spacing	Applications	Remarks
864–865	10 mW	50 kHz	Wireless audio applications	no restriction
868–868.6[4]	25 mW	[6]	Non-specific use	0.1% LDC, or LBT + AFA[3]
868.6–868.7	10 mW	25 kHz The whole band may be used as 1 wideband channel for high-speed data transmission	Alarm	1% LDC
868.7–869.2[4]	25 mW	[6]	Non-specific use	0.1% LDC, or LBT + AFA[3]
867.2–867.25	10 mW	25 kHz	Social alarms	0.1%
867.25–869.3	10 mW	25 kHz	Alarms	0.1%
869.3–869.4	10 mW	25 kHz	Alarms	1%
869.4–869.65	500 mW	≤ 25 kHz The whole band may be used as 1 wideband channel for high-speed data transmission	Non-specific use	10% LDC, or LBT + AFA[3]
869.65–869.7	25 mW	25 kHz	Alarms	10%
869.7–870[5]	25 mW	No restriction	Non-specific use	1% LDC, or LBT + AFA[2,3]
869.7–870[5]	5 mW	No restriction	Non-specific use	No restriction

Notes [1–9] in Tables 2.16 and 2.17 are given as follows.

[1] The power limits, channel arrangement, and duty cycle for frequency hopping spread spectrum (FHSS) equipment, direct sequence spread spectrum (DSSS), and other non-FHSS spread spectrum equipment are described in detail in ETSI EN 300 220-1.

[2] For frequency agile devices without LBT (or equivalent techniques) operating in the frequency range 863–870 MHz, the duty cycle limit applies to the total transmission unless specifically stated otherwise.

[3] When either a duty cycle, LBT, or equivalent technique applies then it shall not be user dependent/adjustable and shall be guaranteed by appropriate technical means. For LBT devices without adaptive frequency agility (AFA) or equivalent techniques, the duty cycle limit applies.

[4] Devices supporting audio and video applications shall use a digital modulation method with a maximum bandwidth of 300 kHz. Devices supporting analogue and/or digital voice shall have a maximum bandwidth not exceeding 25 kHz.

[5] Devices shall not support audio and/or video applications. Devices supporting voice applications shall not exceed 25 kHz bandwidth and shall use spectrum access technique such as LBT or equivalent; the transmitter shall include a power output sensor controlling the transmitter to a maximum transmit period of 1 minute for each transmission.

[6] The preferred channel spacing is 100 kHz allowing for subdivision into 50 kHz or 25 kHz.

[7] The power density can be increased to +6.2 dBm/100 kHz and −0.8 dBm/100 kHz, if the band is limited to 865–868 MHz and 865–870 MHz, respectively.

[8] For wideband modulation other than FHSS and DSSS with a bandwidth of 200 kHz to 3 MHz, duty cycle can be increased to 1% if the band is limited to 865–868 MHz and power to ≤ 10 mW ERP.

[9] Duty cycle may be increased to 1% if the band is limited to 865–868 MHz.

different countries, it is desirable to designate a worldwide frequency band to support for mobility. Moreover, WBAN may use spectrum with high user density, it is desirable to designate frequency bands to WBAN with large available bandwidths.

As shown in Figure 2.6, the first worldwide frequency band is 402–405 MHz, which is limited to MICS commonly. Although EU and USA have extended the original frequency band with two wings of 401–402 MHz and 405–406 MHz, the common available bandwidth worldwide is only 3 MHz. This bandwidth is enough to support communication for most of the implant MICS devices including defibrillator, pacemaker, nerve stimulators, and so on. However, it is difficult to provide data rate of up to 10 Mbps, which is required to transmit real-time picture from a capsule endoscope.

The second worldwide frequency band is 2.4 GHz ISM band. Although ITU-RR designates 1 GHz bandwidth, a number of wireless systems, including WLAN and Bluetooth, have occupied the spectrums. As a result, spectrum available for WBAN will be limited. High data rate transmission up to 10 Mbps is not easy at this band. Coexistence between different wireless systems at the same band must be well studied.

The third wide recognized frequency band is UWB. Under the current regulations, 4.2–4.8 GHz can be used without DAA or LDC before the end of 2010. However, Japan and Korea are in pre-decision to extend the time limit further for three years or even more. Moreover, 7.25–8.5 GHz can be used without specific restrictions in all available regulations. Therefore, UWB has a strong potential to provide high data rate transmission up to 10 Mbps. The problem for UWB is that in most regulations except for USA, UWB is only allowed to indoor usage, although EU allows UWB to be used for road or rail vehicles. This restriction may deter manufacturers and users who want to enable UWB 24 hours anywhere.

References

[1] http://www.itu.int/en/pages/default.aspx.
[2] http://www.fcc.gov/.
[3] http://ec.europa.eu/index_en.htm.
[4] http://www.miit.gov.cn/.
[5] http://www.soumu.go.jp/english/index.html.

[6] http://www.icnirp.de/.

[7] http://webstore.iec.ch/webstore/webstore.nsf/artnum/033746.

[8] IEEE Engineering in Medicine and Biology Society, "Health informatics — PoC medical device communication; Part 00101: Guide — Guidelines for the use of RF wireless," IEEE Std 11073-00101™-2008, New York, December 2008.

[9] "First Report and Order: Revision of Part 15 of the Commission's Rules Regarding Ultra-Wideband Transmission Systems," ET Docket 98–153, FCC 02-48, April 2002. (http://hraunfoss.fcc.gov/edocs_public/attachmatch/FCC-02-48A1.pdf).

[10] "Commission Decision of 21 April 2009 — amending Decision 2007/131/EC on allowing the use of the radio spectrum for equipment using ultra-wideband technology in a harmonised manner in the Community," 2009/343/EC, April 2009.

[11] CEPT Report 34, "Report B from CEPT to European Commission in response to the Mandate 4 on Ultra-Wideband (UWB)," October 2009.

[12] Draft ETSI EN 302 065 V1.2.1, "Electromagnetic compatibility and Radio spectrum Matters (ERM); Short Range Devices (SRD) using Ultra Wide Band technology (UWB) for communications purposes; Harmonized EN covering essential requirements of article 3.2 of the R&TTE Directive," December 2009.

[13] D. Rosener, *Introduction to Bluetooth Engineering*. John Wiley & Sons Inc., 2007.

[14] F. Eady, "Hands-On Zigbee: Implementing 802.15.4 With Microcontrollers," Newnes, 2007.

[15] ITU-R SA.1346, "Sharing between the meteorological aids service and medical implant communication systems (MICS) operating in the mobile service in the frequency band 401–406 MHz," 1998.

[16] FCC-06-103A1, "Notice of proposed rulemaking notice of inquiry and order," July 2006.

[17] ERC/DEC/(01)17, "ERC Decision of 12 March 2001 on harmonised frequencies, technical characteristics and exemption from individual licensing of Short Range Devices used for Ultra Low Power Active Medical Implants operating in the frequency band 402–405 MHz, " March 2001.

[18] ETSI EN 301 839-1 V1.3.1, "Electromagnetic compatibility and Radio spectrum Matters (ERM) Short Range Devices (SRD); Ultra Low Power Active Medical Implants (ULP-AMI) and Peripherals (ULP-AMI-P) operating in the frequency range 402 MHz to 405 MHz; Part 1: Technical characteristics and test methods," January 2009.

[19] ETSI EN 302 537-1 V1.1.2, "Electromagnetic compatibility and Radio spectrum Matters (ERM); Short Range Devices (SRD); Ultra Low Power Medical Data Service Systems operating in the frequency range 401 MHz to 402 MHz and 405 MHz to 406 MHz; Part 1: Technical characteristics and test methods," December 2007.

[20] APT/AWF, "APT recommendation on spectrum use and sharing by very low power wireless heart implant transmitters," November 2007.

[21] FCC 09-57, "Notice of proposed rulemaking," June 2009. (http://hraunfoss.fcc.gov/edocs_public/attachmatch/FCC-09-57A1.pdf).

[22] ETSI TR 102 655 V1.1.1, "Electromagnetic compatibility and Radio spectrum Matters (ERM); System reference document; Short Range Devices (SRD); Low Power Active Medical Implants (LP-AMI) operating in a 20 MHz band within 2 360 MHz to 3 400 MHz," November 2008.

[23] http://www.acma.gov.au/WEB/STANDARD/pc=PC_2625.

[24] http://www.rsm.govt.nz/cms/policy-and-planning/current-projects/radiocommunications/ review-of-spectrum-allocations-for-short-range-devices/an-engineering-discussion- paper-on-spectrum-allocations-for-short-range-devices/6-medical-telemetrybiomedical telemetry transmitter.

[25] ETSI EN 300 220-1 V2.3.1, "Electromagnetic compatibility and Radio spectrum Matters (ERM); Short Range Devices (SRD); Radio equipment to be used in the 25 MHz to 1000 MHz frequency range with power levels ranging up to 500 mW; Part 1: Technical characteristics and test methods," February 2010.

3

Antenna, Body Tissues and Radio Propagation

3.1 Introduction

Wearable and implantable electronic devices are becoming very popular in personal communications and wireless monitoring of vital functions. Therefore, there is an increasing demand for effective communication technologies to support emerging health care delivery systems, hence, wireless body area network (WBAN) for sensing and monitoring of vital signs is the one of most rapid growing wireless communication system.

For wearable or on-body communications, the main design criteria for WBAN units are physical size, weight, and power efficiency, all of which are directly related to antenna performance. Therefore, a key component of wireless body area network is an antenna. It must meet biocompatible, size limit requirements, and faces numerous radio frequency (RF) challenges, such as suitable radiation characteristics for on-body communications and with relatively high radiation efficiency, and consideration of absorbed power by body tissues.

There is huge interest in WBAN antennas, both on in-body and on-body, and also both on narrowband and wideband. Research on WBAN antennas has rapidly increased in recent years [1–9].

In radio communications, the most critical obstacle is always received power. Radio communications become technologically challenging whenever human body is involved in the communication links, due to safety issues and power limit. Another challenge related with wearable antennas is the pattern degradation caused by the irregular body shape or clothing.

35

Almost all antenna designers are struggling with high dielectric of body, which has serious effect on wearable antenna. Therefore, some designers like to use ground plane to shield the body tissues from antenna to increase antenna gain and efficiency. However, one of the main requirements for WBAN antenna is small size and published papers are not showing any significant effect of ground plane on antenna gain. Therefore, for miniaturization of antenna it would be desirable to eliminate the ground plane. Obviously in designing antenna the electromagnetic interaction among on-body units and the human body is an important factor to be considered. The results of inappropriate antenna design may have significantly effect on transmission range and quality of transmission line.

Electromagnetic energy can be transmitted into body tissues by antennas. The behavior of electromagnetic fields and their coupling and interaction with body tissues are very different, depending on distance, angle, near-field or far-field, frequency of operation, and the physical dimension of tissue.

An important step in the development of a WBAN is the characterization of the electromagnetic wave propagation from devices that are close to the human body. Another important consideration for WBAN applications is the ability to efficiently couple two low-profile, compact nodes that do not have line-of-sight. Optimization of the link efficiency and path loss must be quantified for reason of expected radiation performance and link budget calculation due to effect of body tissues on output power and radiation pattern shape.

3.2   Type of Antenna

Most of the medical wireless applications for the human body are involved in the electromagnetic coupling into and/or out of the human body. This coupling usually requires an antenna to transmit a signal into a body or pick up a signal from a body. The antenna operating environment for WBAN is different from the traditional free-space communications. The WBAN antennas may be classified into two main groups: (1) electrical antennas, such as dipoles and horns; and (2) Magnetic antennas, for instance loops and slots.

Electric antennas are characterized by strong electric fields close to the antenna, whereas, magnetic antennas are distinguished by strong magnetic fields close to the antenna. Electric antennas are more prone to couple to nearby objects than magnetic antennas. Thus, magnetic antennas are preferred for applications involving embedded antennas.

Some antennas such as helical-coil, are similar to magnetic antennas in some respect, but their heating characteristics appear to be more like that of electrical antennas. The strong E-field generated between the turns of the coil is mainly responsible for tissue heating.

The general view of the electrical and magnetic antennas shows that the small magnetic antennas would be better candidate for WBAN device than electrical antennas in terms of the radiation performance and safety issues.

3.2.1 Electrical Antenna

Electrically small antennas are antennas with geometrical dimensions which are small compared to the wavelengths of the radiated fields. This type of antenna typically generates large components of E-field normal to the tissues interface, which overheat the body tissues. This is because boundary conditions require the normal E-field at the interface to be discontinuous by the ratio of the permittivities, and since fat has a lower permittivity than muscle, the E-field in the fat tissue is higher [10].

3.2.2 Magnetic Antenna

The magnetic counterpart of the electrical dipole antenna is a current loop with small radius compared with the wavelength. Such current loop is equivalent to a magnetic dipole along the axis normal to the plane of the loop [11].

This type of antenna produces an E-field mostly tangential to the tissues, which seem not to couple as strongly to the body as electrical antennas. Therefore, it does not overheat the body tissues.

3.2.3 Antenna Design

No doubt that antennas are an essential part of any wireless communication system, and printed dipole and monopole antennas are comprehensively used in diverse wireless applications due to their many advantages, such as low cost, easy to fabricate, light weight, and low profile [12, 13]. It was pointed out earlier that electrical antenna typically generates large components of

E-field normal to the tissues, which overheat the tissue. However, the magnetic antenna produces an E-field mostly tangential to the tissues, which seem not to couple as strongly to the body as electrical antennas. Therefore, it does not overheat the tissue. The dipole and loop are two basic antennas with opposite radiation patterns.

An antenna can be designed in either air or the dielectric of the body. If the antenna is designed in air, the antenna's best performance will be achieved when air surrounds the antenna. If an antenna is designed in or near the dielectric of the body, the best performance from the antenna will be achieved when the antenna is actually inside the body cavity or placed on body surface. Therefore, to design an antenna for WBAN, it is necessary to place the antenna in the medium in which it will be expected to operate. From a connection point of view, the antenna appears to be a dual gate. The connection which is not made to RF-cable is connected to the environment, therefore one must always note, that the surroundings of the antenna have a strong influence on the antenna's electrical features.

The main design criteria for WBAN antennas are physical size. However, with size reduction at a fixed operation frequency, the impedance bandwidth of a printed antenna is usually decreased [14]. Moreover, when the antenna size is reduced at a fixed operating frequency, the antenna gain is also decreased. In order to give a clear understanding of WBAN antenna design we will present an ultra-wideband (UWB) antenna design feature for free space and on-body tissues.

3.2.3.1 Free-space antenna

The loop antennas can be used for the UWB WBAN communications. However, a conventional wire loop antenna shows less than 10% bandwidth for a 2:1 voltage standing wave ratio (VSWR). Therefore, conventional loop antenna went under different modifications to increase the bandwidth. A broadband loop antenna introduced in [15] has a small gap in the wire loop. This small gap increases the impedance bandwidth to more than 24%.

In this design we present a loop antenna whose left and upper arms together introduce an L-shape [16]. However, the L-shaped antenna itself is a class of broadband planar antenna, which allows the broad impedance bandwidth and less cross-polarization radiation [17, 18].

The structure of L-loop antenna is illustrated in Figure 3.1. To have a linearly polarized radiation the total length of outer limits of the square loop antenna should be in one wavelength [19]. Designing an antenna for 3.1 GHz will give the wavelength of $\lambda_0 = 96.77$ mm. The present antenna is composed of a single metallic layer and is printed on a side of a flame retardant 4 (FR4) substrate with dielectric constant of $\varepsilon_r = 4.4$, loss tangent of $\tan \delta = 0.02$, and thickness of 1 mm. A coupled tapered transmission line is printed in the same side with similar metallic layer. A copper of 0.018 mm thickness has been used as a metallic layer. As shown in Figure 3.1, the size of the proposed

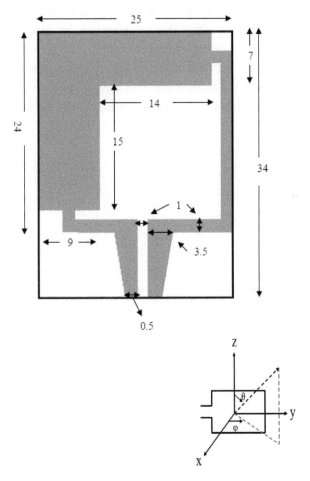

Fig. 3.1 Geometry of antenna with dimensions in millimeters.

antenna is 24 × 25 × 1 mm, which is quite appropriate for wireless systems. The rectangular loop has 98 mm length, which is fairly close to one wavelength of designed antenna. In this work we used taper transmission line for impedance matching, and we modified the shape of conventional loop antenna with introducing an L portion to its arms, as shown in Figure 3.1 [7], to reduce the antenna internal reflections at its discontinuities and make gradual transition between the metal surface of the antenna and free space. The broadband characteristics of the proposed antenna can be understood by its current distribution as the current is relatively more distributed over the antenna L portion, Figure 3.2. The antenna prototype is illustrated in Figure 3.3.

The tapered transmission lines have shown good impedance matching over a wide range of frequency [20–25]. The antenna is fed from a 50 Ω connector through a coupled tapered transmission line. The geometry of the taper is chosen to minimize the reflection and optimize impedance matching and bandwidth. A tapered structure in the feed area creates low current standing wave ratios, therefore more magnitude of pulse near to the feed point.

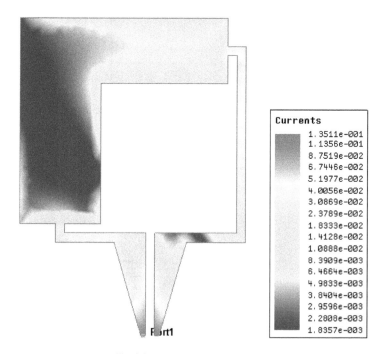

Fig. 3.2 Antenna current distribution.

Fig. 3.3 Photo of the antenna prototype.

In UWB communications, the antennas are significant pulse-shaping filters. Any distortion of the signal in the frequency domain causes distortion of the transmitted pulse shape, therefore, increasing the complexity of the detection mechanism at the receiver. UWB antennas require the phase center and the VSWR to be constant across the whole bandwidth of operation. A change in phase center may cause distortion on the transmitted pulse and worse performance at the receiver [26]. Magnetic antennas may be considered as current-driven elements and have mainly magnetic near field. Since electric fields tend to couple more strongly to nearby objects, magnetic antennas are better suited for embedded applications [27] in particular for WBAN applications, where human body is involved to the communication links.

The achieved impedance bandwidth is in the order of 2 GHz (3.1–5.1 GHz) for VSWR ≤ 2, as illustrated in Figure 3.4.

Radiation patterns of the UWB loop antenna in the $y - z$ plane and $x - z$ planes at frequencies of 3.1, 4.1, and 5.1 GHz for $\varphi = 0^o$ and $\varphi = 90^o$, for the free space, are shown in Figures 3.5 and 3.6. The results demonstrate that the radiation patterns are quite similar in the three different frequencies, which is a very important factor for the wireless system with high data rate and wide bandwidth.

For the UWB radios, an antenna which provides linear phase is needed. The phase linearity is directly related to the group delay of the antenna. The phase response and group delay are related to the antenna gain response. The group delay variation induced by the radiation pattern of the antenna appears

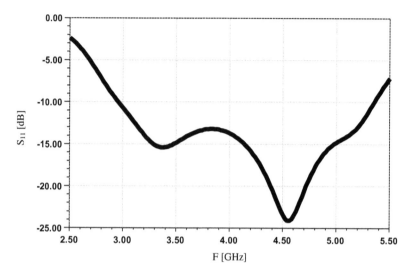

Fig. 3.4 Return loss of the UWB loop antenna.

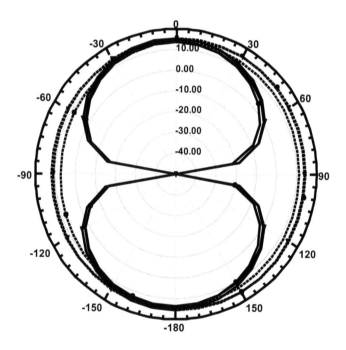

Fig. 3.5 Radiation patterns in $y - z$ plane of the UWB loop antenna for —— $\varphi = 0$ and —— $\varphi = 90$ at 3.1 GHz, ▼4.1 GHz, and ●5.1 GHz.

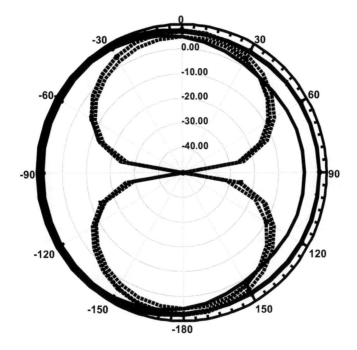

Fig. 3.6 Radiation patterns in $x-z$ plane of the UWB loop antenna for —— $\varphi = 0$ and —— $\varphi = 90$ at 3.1 GHz, ▼4.1 GHz, and •5.1 GHz.

to be a very important parameter in the overall receiver system performance, since it can bring relatively large timing errors.

An antenna in UWB system can be analyzed as a filter by means of magnitude and phase responses. When a signal passes through a filter, it experiences both amplitude and phase distortion, depending on the characteristics of the filter. By representing the receiver/transmitter antenna as a filter, we can determine its phase linearity within the frequency band of interest by looking at its group delay. Group delay is the measure of a signal transition time through a device. It is classically defined as the negative derivative of phase versus frequency given by:

$$GroupDelay = -\frac{d\theta(\omega)}{d\omega},$$ (3.1)

If there is a null in the antenna gain, it implies a nonlinear phase, and therefore, a non-constant group delay. A non-constant group delay will distort the

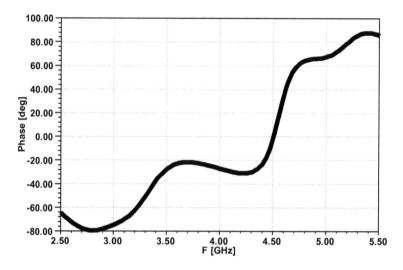

Fig. 3.7 Phase response of the UWB loop antenna in free space.

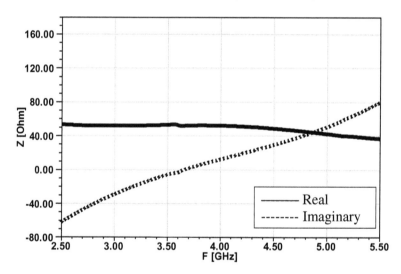

Fig. 3.8 Impedance of the UWB loop antenna in free space.

antenna impulse response. The phase response of the loop antenna is shown in Figure 3.7. It shows approximately linear phase response.

Figure 3.8 illustrates antenna input impedance for both real and imaginary parts. The results indicate that the impedance behavior is almost linear and constant around 50 Ω.

3.2.3.2   Free-space antenna placed on body tissue

The electromagnetic interaction between humans and wireless device antennas has been an important topic in the past decades. Initially, all the studies were focused for handsets and mostly on human head [28–31] But nowadays, the human and wireless device interaction has to be studied in a wide range of scenarios, ranging from the on-body to the in-body wireless communication.

The influence of the human tissues on antenna characteristics will be appear by altering antenna input impedance, modifying the antenna radiation pattern, absorbing antenna delivered power, and so on.

A model of three-layer tissue of a human body composed of skin, fat, and muscle has been developed to study the effects of human body on antenna characteristics as shown in Figure 3.9. It has dimensions of $134 \times 125 \times 33$ mm. The total thickness is 33 mm, which consists of three layers as, skin 1 mm, fat 2 mm, and muscle 30 mm. The effect of blood vessels, sweat gland, and any form of water-content tissues is not considered in this simulation. The thickness of skin in human body varies from 0.5 mm on eyelids to 4 mm or more on the palms of hands and the soles of feet. The thickness of fat and muscle will vary with respect to gender and body structures.

Figure 3.10 shows the antenna return loss in the proximity of tissue layers. The impedance bandwidth is dropped down to 1 GHz and the lower frequency is shifted to 2.6 GHz. Input impedance of the UWB antenna in the proximity of body tissue is illustrated in Figure 3.11.

Radiation patterns of the UWB loop antenna in the $y-z$ and $x-z$ planes at frequencies of 3.1, 4.1, and 5.1 GHz for $\varphi = 0^o$ and $\varphi = 90^o$, for the proximity to human body, are shown in Figures 3.12 and 3.13, respectively.

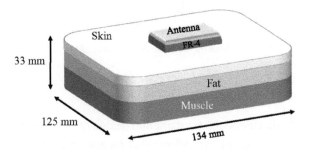

Fig. 3.9 Body tissue model with antenna on top of skin.

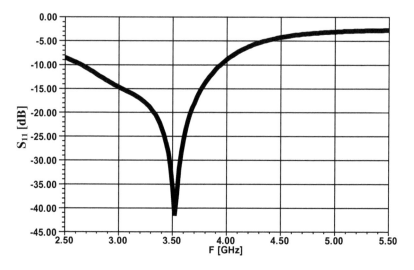

Fig. 3.10 Return loss of the UWB loop antenna on body surface.

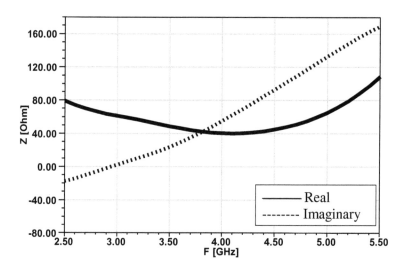

Fig. 3.11 Input impedance in the proximity of body tissue.

3.2.3.3   On-body antenna

Having already discussed the fundamental design of WBAN antenna, our goal is to illustrate some simple guidelines that an antenna could be used for free space and on-body applications.

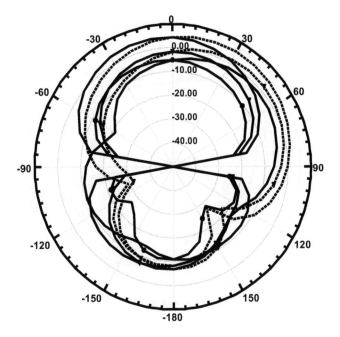

Fig. 3.12 Radiation patterns in $y - z$ plane of the UWB loop antenna in the proximity to human body for ------ $\varphi = 0$ and —— $\varphi = 90$ at 3.1 GHz, ▼4.1 GHz, and ●5.1 GHz.

In order that an antenna to be suitable for wearable body area communication, it should be placed near to the body area with a sufficient gap between an antenna and human body. The simulation results showed that a minimum separation of 14 mm is necessary to reduce the effect of the human body on the antenna characteristics. Therefore, in our work we added a second layer substrate, i.e., RH-5 with thickness of 14 mm, $\varepsilon_r = 1.09$ and $\tan \delta = 0.0004$, as shown in Figure 3.14. This substrate is chosen due to its relative permittivity, which is near to air and it won't affect the antenna parameters. Therefore, same antenna could be used for on-body to on-body communication and on-body to off-body as well.

The return loss of the antenna in proximity of human body with RH-5 substrate layer is illustrated in Figure 3.15. It can be seen that same level of return loss of free space is achieved.

Also, input impedance shows constant 50 Ω around the band, as shown in Figure 3.16. A well-matched UWB antenna will have a smooth and continuous impedance transformation. This impedance transition makes the antenna to

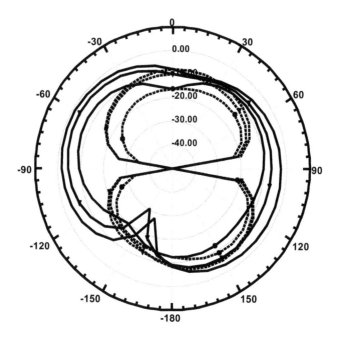

Fig. 3.13 Radiation patterns in $x - z$ plane of the UWB loop antenna in the proximity to human body for
———· $\varphi = 0$ and —— $\varphi = 90$ at 3.1 GHz, ▼4.1 GHz, and •5.1 GHz.

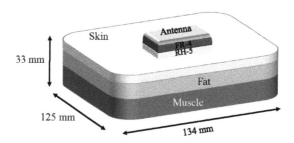

Fig. 3.14 Antenna with RH-5 substrate on body tissue model.

radiate efficiently and provide an excellent match to a radio frequency (RF) front end.

Radiation patterns are shown in Figure 3.17 for $y - z$ plane and Figure 3.18 for $x - z$ plane for the frequencies of 3.1, 4.1, and 5.1 GHz. Figures 3.17 and 3.18 show that the radiation patterns are having a very good agreement with the free-space radiation patterns.

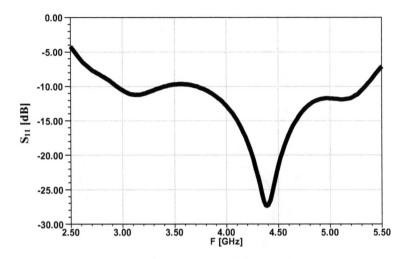

Fig. 3.15 Return loss of the UWB loop antenna on body surface with second substrate.

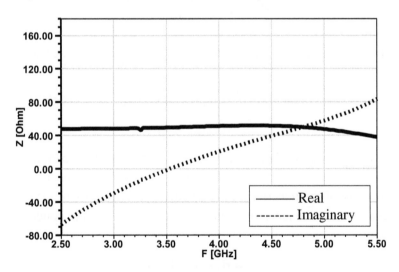

Fig. 3.16 Input impedance in the proximity of body tissue with RH-5 substrate.

3.3 Electromagnetic Radiation and Human Tissues

The dielectric properties of materials vary at different frequencies and are generally defined in terms of their relative permittivity and conductivity. Changes in the dielectric properties of materials can be due to several factors such as frequency, temperature, and applied electric field. The dielectric properties are

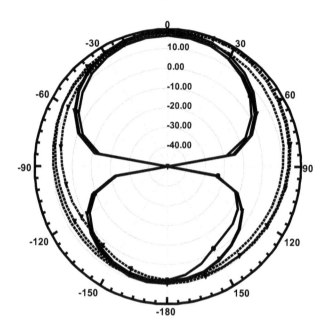

Fig. 3.17 Radiation patterns in $y - z$ plane of the UWB loop antenna in the proximity to human body with RH-5 substrate for ┄┄• $\varphi = 0$ and ── $\varphi = 90$ at 3.1 GHz, ▼4.1 GHz, and •5.1 GHz.

normally defined in the form of complex permittivity (ε^*) and given by:

$$\varepsilon^* = \varepsilon_0(\varepsilon' - j\varepsilon''), \tag{3.2}$$

where ε_0 is the permittivity of free space, ε' is the relative permittivity and ε'' is the dielectric loss or loss factor. The loss factor is associated with the conductivity of the material [32] and material conductivity (σ) is related to the displacement conductivity, but in biological tissues the conductivity is given by:

$$\sigma = \sigma_d + \sigma_i \quad \text{(S/m)}, \tag{3.3}$$

where σ_d is displacement conductivity and σ_i is the ionic conductivity because of the drift of free ions due to field in biological tissue [33]. Dielectric loss represents the ionic conductivity and the absorption due to processes of relaxation for biological tissues.

The frequency dependence of ε and σ is directly related to the polarization of molecules and structural interfaces caused by an applied electric field within the biological tissue. The complex physical nature of biological tissues

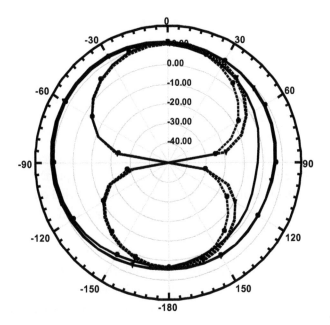

Fig. 3.18 Radiation patterns in $x - z$ plane of the UWB loop antenna in the proximity to human body with RH-5 substrate for ------· $\varphi = 0$ and —— $\varphi = 90$ at 3.1 GHz, ▼4.1 GHz, and ●5.1 GHz.

allows several relaxation processes to take place simultaneously. Hence, the total electrical response of the material can be characterized by several time constants. The first order Debye equation is given by [34]:

$$\varepsilon^*_{(\omega)} = \varepsilon_\infty + \frac{\varepsilon_s - \varepsilon_\infty}{1 + j\omega\tau}, \tag{3.4}$$

where ε^* is the complex permittivity, ε_s is the relative permittivity at low frequencies, and ε_∞ corresponds to the permittivity at high frequencies.

Furthermore, Sullivan considered a Debye model with an added conductivity term as an approximate model for muscle tissue [35]:

$$\varepsilon^*_{(\omega)} = \varepsilon_\infty + \frac{\varepsilon_s - \varepsilon_\infty}{1 + j\omega\tau} + \frac{\sigma}{j\omega\varepsilon_0}, \tag{3.5}$$

3.3.1 Electromagnetic Properties of Human Body Tissues

The electrical properties of human tissues have been of interest for many reasons over past few decades. Gildemeister (1928), Einthoven and Bijtel (1923), and Hozawa (1932) measured the resistance and capacity of human skin in the audio-frequency range [36]. One of the initial extensive works on human tissues and organs in the frequency range of 10 MHz–100 GHz was made by Durney et al. [33]. However, its first edition (September 1976) only covered the frequency range of 10 KHz–1.5 GHz.

The human body consists of materials of different dielectric constants, thickness, and characteristic impedance. Therefore, depending on the frequency of operation, the human body can lead to high losses caused by power absorption, central frequency shift, and radiation pattern destruction. The absorption effects vary in magnitude with both frequency of applied field and the characteristics of the tissue, which is largely based on water and ionic content. Therefore, body tissues are not an ideal medium for radio-frequency wave transmitting.

Biological tissues can be classified into two main categories. Tissues with high amount of water, hence have higher dielectric constant and conductivity such as skin, muscle, kidney and liver, and tissues with low amount of water which have lower dielectric constant and conductivity such as fat and bone. Tissues such as bone marrow, lung, and brain contain intermediate amounts of water and have dielectric constant and conductivities that fall between the values of the other two groups [37]. Unlike the standard communication all the way through constant air, various tissues and organs within the body have their own unique electrical properties. Tables 3.1 and 3.2 show relative permittivity, conductivity, loss tangent, and penetration depth of different tissues at frequencies of 2.4 GHz and 5 GHz [38].

Understanding the human body's effect on RF wave propagation is complicated due to the fact that the body consists of components each of which offers different degrees, and in some cases, different types of RF interactions. The liquid nature of most body structures gives a degree of RF attenuation, whereas the skeletal structure introduces wave diffraction and refraction at certain frequencies.

When the wavelength of a signal is significantly larger than the cross-section of the human body being penetrated, there is very little effect on the signal. These wavelengths occur at frequencies below 4 MHz. Above 4 MHz, the absorption of RF energy increases and the human body may be considered to be essentially dense until roughly 1 GHz when the dielectric properties of the human body begin to introduce a scattering effect on the RF signal. Differences in relative permittivity and conductivity of muscle and fat across a wide range of frequency are shown in Figure 3.19 [39].

Table 3.1. Electrical Properties of Human Body Tissues at 2.4 GHz.

Tissue name	Conductivity (S/m)	Relative permittivity	Loss tangent	Penetration depth (m)
Aorta	1.4041	42.593	0.2496	0.02486
Bladder	0.6726	18.026	0.2794	0.03382
Blood	2.5024	58.347	0.3212	0.01640
Blood vessel	1.4041	42.593	0.2469	0.02486
Bone cancellous	0.7876	18.606	0.3170	0.02942
Bone cortical	0.3845	11.410	0.2524	0.04699
Bone marrow	0.0928	5.3024	0.1311	0.13196
Brain grey matter	1.7730	48.994	0.2710	0.02114
Brain white matter	1.1899	36.226	0.2460	0.02705
Breast fat	0.1334	51.563	0.1938	0.09076
Cartilage	1.7172	38.878	0.3308	0.01953
Cerebrospinal fluid	3.4122	66.319	0.3853	0.01289
Cervix	1.6933	47.673	0.2660	0.02183
Colon	1.9997	53.969	0.2775	0.01968
Cornea	2.2588	51.697	0.3272	0.01711
Eye sclera	1.9967	52.698	0.2837	0.01949
Fat	0.1023	52.853	0.1450	0.11956
Gland	1.9283	57.272	0.2521	0.02099
Heart	2.2159	54.918	0.3022	0.01795
Kidney	2.3901	52.856	0.3386	0.01637
Liver	1.6534	43.118	0.2872	0.02129
Lung deflated	1.6486	48.454	0.2548	0.02259
Lung inflated	0.7902	20.510	0.2885	0.03073
Lymph	1.9283	57.272	0.2521	0.02099
Muscle	1.7050	52.791	0.2419	0.02278
Pancreas	1.9283	57.272	0.2521	0.02099
Prostate	2.1273	57.629	0.2764	0.01912
Skin dry	1.4407	38.063	0.2835	0.02295
Skin wet	1.5618	42.923	0.2725	0.02247
Small intestine	3.1335	54.527	0.4304	0.01278
Spinal cord	1.0681	30.196	0.2649	0.02754
Spleen	2.2000	52.546	0.3135	0.01770
Stomach	2.1671	62.239	0.2607	0.01948
Tendon	1.6440	43.210	0.2849	0.02143
Thyroid	1.9283	57.272	0.2521	0.02099
Tongue	1.7662	52.698	0.2510	0.02198
Uterus	2.2058	57.897	0.2853	0.01849

3.3.2 Computational Methods

Computational methods for electromagnetics are an essential resource for making efficient and perfect formulations for electromagnetics applications and their numerical treatment. Maxwell's equations are the foundation of all electromagnetics. The propagation of electromagnetic is governed by

Table 3.2.   Electrical Properties of Human Body Tissues at 5 GHz.

Tissue name	Conductivity (S/m)	Relative permittivity	Loss tangent	Penetration depth (m)
Aorta	3.5330	39.295	0.3232	0.009538
Bladder	1.5308	16.674	0.3300	0.014347
Blood	5.3951	53.950	0.3595	0.007339
Blood vessel	3.5330	39.295	0.3232	0.009538
Bone cancellous	1.8116	16.050	0.4057	0.011970
Bone cortical	0.9622	10.040	0.3445	0.017731
Bone marrow	0.2337	5.0379	0.1668	0.051143
Brain grey matter	4.0995	45.147	0.3264	0.008813
Brain white matter	2.8588	33.444	0.3073	0.010862
Breast fat	0.3498	4.6450	0.2707	0.033001
Cartilage	4.0855	33.625	0.4368	0.007704
Cerebrospinal fluid	6.5969	61.952	0.3828	0.006445
Cervix	3.9348	44.416	0.3184	0.009102
Colon	4.5845	49.723	0.33147	0.008274
Cornea	4.7223	47.733	0.35567	0.007885
Eye sclera	4.4985	48.996	0.33008	0.008369
Fat	0.2422	5.0291	0.17315	0.049333
Gland	4.6614	53.342	0.31416	0.008417
Heart	4.8626	50.274	0.34772	0.007853
Kidney	4.9423	48.059	0.36971	0.007568
Liver	3.8278	39.260	0.35052	0.008818
Lung deflated	3.9413	44.859	0.31587	0.009130
Lung inflated	1.7220	18.966	0.32641	0.013599
Lymph	4.6614	53.342	0.31416	0.008417
Muscle	4.0448	49.540	0.29353	0.009334
Pancreas	4.6614	53.342	0.31416	0.008047
Prostate	4.8833	53.526	0.32799	0.008057
Skin dry	3.0608	35.774	0.30760	0.010493
Skin wet	3.5744	39.611	0.32441	0.009466
Small intestine	5.7533	49.977	0.41386	0.006655
Spinal cord	2.4260	27.890	0.32171	0.011694
Spleen	4.7178	48.195	0.35193	0.007928
Stomach	5.1565	57.890	0.32023	0.007930
Tendon	4.3009	38.299	0.40372	0.007787
Thyroid	4.6614	53.342	0.31416	0.008417
Tongue	4.2680	48.996	0.31317	0.008810
Uterus	4.9790	53.681	0.33345	0.007917

Maxwell's equations. Maxwell's equations are as follows:

$$\nabla \times H = J \frac{\partial D}{\partial t}, \tag{3.6}$$

$$\nabla \times E = -\frac{\partial B}{\partial t}, \tag{3.7}$$

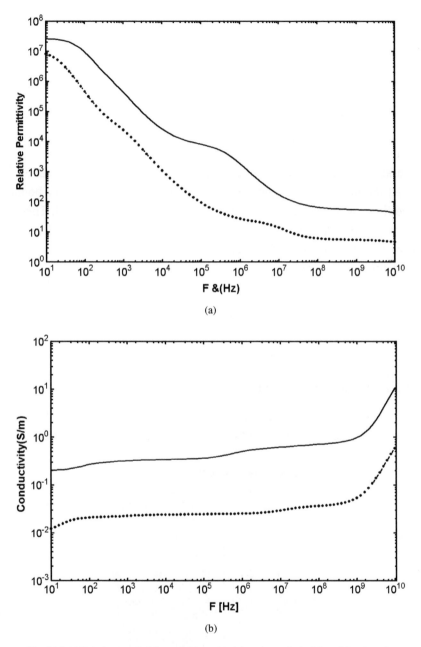

(a)

(b)

Fig. 3.19 (a) Relative permittivity and (b) conductivity of muscle (solid) and fat (dotted).

$$\nabla \bullet D = \rho, \tag{3.8}$$

$$\nabla \bullet B = 0, \tag{3.9}$$

3.3.2.1   Boundary conditions

Electromagnetic waves have their electric and magnetic fields perpendicular to the direction of propagation. The properties of the medium in which the waves propagate are characterized by the two material constants, permittivity (ε) and the permeability (μ), of the medium, as shown in Figure 3.20.

At the interface, the tangential and normal fields must satisfy boundary conditions, which are consequences of Maxwell's equations [40]. These two electric field polarizations have quite different transmission and reflection properties at boundaries. The relations between fields E, H, and the material constants ε and μ are formulated as:

$$\frac{\partial E}{\partial t} = \frac{1}{\varepsilon}\nabla \times H - \frac{\sigma}{\varepsilon}E, \tag{3.10}$$

$$\frac{\partial H}{\partial t} = -\frac{1}{\mu}\nabla \times E, \tag{3.11}$$

3.3.2.2   Method of moments

The method of moment (MoM) is a full wave solution of Maxwell's integral equations in the frequency and time domain. The MoM is theoretically quite

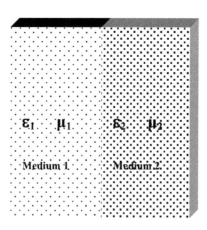

Fig. 3.20 The boundary between two materials with different properties.

straightforward. It basically utilizes the equations of unknown fields in integral form to determine the field distribution in a given medium [41]. The MoM requires calculating only boundary values, rather than values throughout the space defined by a partial differential equation, it is significantly more efficient in terms of computational resources for problems where there is a small surface/volume ratio. They can be used to solve closed or open boundaries and do not need truncation of the perfect metallic layers (PML) or absorbing boundary conditions (ABC).

3.3.2.3 Finite element method

The finite element method (FEM) is a numerical technique for solving models in differential form. The FEM has its formal basis in the Galerkin procedure of weighted residuals. For a given design, the FEM requires the entire structure or problem, including the surrounding regions, to be modeled with finite number of smaller regions. Finite elements is breaking up a problem into small regions and solutions are found for each region taking into account only the regions that are right next to the one being solved. In the case of magnetic fields where FEM is often used, the vector potential is what is solved for in these regions. Magnetic field solutions are derived from the vector potential through differentiating the solution. This can cause problems in smoothness of field solutions. Theoretically, any partial differential equation class of problem can be solved using FEM [42].

3.3.2.4 Finite difference time domain

The finite difference time domain (FDTD) solves Maxwell's equations in the time domain. Because of this, the calculation of the electromagnetic field values progresses at discrete steps in time. Therefore, it gives broadband output from a single execution of the program.

The basic FDTD space grid and time-stepping algorithm trace back to 1966 paper by Yee [43]. In this approach, both space and time are divided into discrete segments or cells, which are small compared to the wavelength. The electric fields are located on the edges of the box and the magnetic fields are positioned on the faces as shown in Figure 3.21. Cell size, the dimensions of the box, is the most important constraint in any FDTD simulation since it

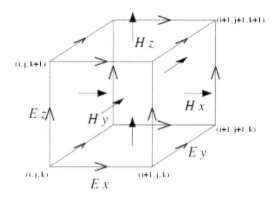

Fig. 3.21 The Yee cell.

determines not only the step size in time, but also the upper frequency limit for the calculation.

FDTD has also been identified as the preferred method for performing electromagnetic simulations for biological effects from wireless devices [44]. The FDTD method has been shown to be the most efficient approach and provides accurate results of the field penetration into biological tissues.

The limitation of FDTD is that very large computational grids are required for long structure. Moreover, the FDTD requires the entire computational structure to be meshed, and these cells must be small compared to the smallest wavelength, which makes modeling very difficult for thin structures.

3.3.3   Incident Field

The electromagnetic (EM) field is composed of electric (E) and magnetic (H) fields. The interaction of EM fields with body depends upon the field characteristics, radiation frequency, the body size, shape, and the dielectric properties of the material under exposure. The maximum coupling of EM field can be occurred when the long axis of the human body is in the direction of the electric field and while the size of human body is of the same order of scale of the wavelength.

Tissues exposed to EM field by a WBAN antenna might be subject to both thermal and non-thermal biological effects. There are severals mechanisms which could be used to explain these effects, such as electrical charges within the biological tissues and magnetic dipoles in molecules. The electrical charges within the body tissues can gain energy due to the forces of applied electric field of EM waves. On the other hand as permeability of biological tissues is close to the free space, the magnetic field of EM waves has less effect on magnetic dipoles within the tissues, hence for less energy will be transferred.

It is evident from above that biological effects from EM fields are much more likely to be caused by electric rather than magnetic fields. Therefore, a magnetic antenna could be suitable candidate for WBAN applications. Due to conductive nature of human tissues any magnetic field close to the body will be tangential to the body. Therefore, if the axis of a magnetic loop antenna is tangential to the body, the maximum magnetic field will be captured, whatever the orientation of the body is [45].

3.3.4 Specific Absorption Rate

Specific absorption rate (SAR) is the unit of measurement for the amount of RF energy absorbed by the body tissues. The SAR is used as a safety standard for irradiation level of electromagnetic wave to the body tissues. Excessive exposure to radio frequency energy can damage human tissue. Accordingly exposure to intense microwaves in excess of 20 mW/cm^2 of body surface is harmful. To evaluate the influence of electromagnetic wave radiation on the body tissue, the SAR is defined as the time derivative of the energy (dW) dissipated in a mass (dm) enclosed in a volume element (dV) of a given tissue density (ρ) [46, 47]:

$$SAR = \frac{d}{dt}\left(\frac{dW}{dm}\right) = \frac{d}{dt}\left(\frac{dW}{\rho dV}\right) \quad \text{(W/kg)}, \qquad (3.12)$$

The SAR is related to the induced electric field by:

$$SAR = \frac{\sigma |E|^2}{\rho} \quad \text{(W/kg)}, \qquad (3.13)$$

where σ is the electrical conductivity of the tissue and E is the induced electric field. A major interaction mechanism is through the currents induced in tissues, so effects are dependent on frequency, wave shape, and intensity. Many of the biological effects of acute exposure to electromagnetic fields are consistent with responses to induced heating, resulting either in rises in tissue temperature or body temperature. The SAR can be classified as; the whole-body SAR, the localized SAR, and short-pulse SAR [48].

The specific absorption (SA) is used as a dosimetric parameter for pulse waveform radiation and is defined as the amount of the energy (*dW*) absorbed by a mass contained in a volume of a given density ($\rho \cdot dV$).

For the safety of human body from the radiation, the amount of radiation is restricted to certain level. Experiments show that exposure to an SAR of 8 W/kg in any gram of tissue in the head or torso for 15 min may have a significant risk of tissue damage [49].

Several organizations have set exposure limits for acceptable radio frequency safety via SAR levels, such as the International Commission on Non-ionizing Radiation Protection (ICNIRP), the Institute of Electrical and Electronic Engineers (IEEE), and the National Radiological Protection Board (NRPB).

SAR is generally quoted as a figure averaged over a volume corresponding to either 1 g or 10 g of body tissue. SAR limits are expressed for two different classes: occupational or controlled exposure and non-occupational or uncontrolled exposure. The limit for uncontrolled exposure is five times more stringent than the limit for the controlled exposure, whose environment and exposure can be monitored and controlled. The limits are defined for exposure of the whole body, and partial body. SAR limits are based on whole-body exposure levels of 0.4 W/kg for occupational and 0.08 W/kg for the non-occupational. Europe, Japan, and Korea have adopted 2 W/kg for 10 g volume averaged SAR. For the United States, the limit is 1.6 W/kg for 1 g volume averaged SAR. The lower U.S. limit is more stringent because it is volume averaged over a smaller amount of tissue [46].

The heating effect of radio frequency radiations destroys living tissue when the temperature of the tissue exceeds 43°C. Therefore, the possible rise of temperature in tissues is very important parameter [50, 51]. The rise of temperature induced by a given SAR distribution associated with an electromagnetic wave emission by thermal exchange via neighboring tissues and the external environment [52, 53] is given by:

$$\frac{dT}{dt} = \frac{SAR}{c} \quad (C°/s), \qquad (3.14)$$

where, *c* is specific heat capacity (J/kg°C), *dt* is duration of exposure, and *dT* is change in temperature. The regulation limits the temperature rise to 1°. This limit is being determined by the specific heat of the tissue. In Equation (3.14), the blood flowing cooling rate, the metabolic heating rate, and the heat losses rate for a tissue have been ignored.

The Bio-heat equation will take into account all the parameters which have been ignored in Equation (3.14) to find the temperature rise due to induced SAR [54].

$$
\rho_{(x,y,z)} C_{(x,y,z)} \frac{\partial T_{(x,y,z,t)}}{\partial t} = \begin{bmatrix} \nabla \bullet (k_{(x,y,z)} \bullet \nabla T_{(x,y,z)}) + h_{m(x,y,z)} + \rho_{(x,y,z)} \\ \bullet SAR_{(x,y,z)} - B_{(x,y,z)} \bullet (T_{x,y,z,t} - T_b) \end{bmatrix}
$$
$$
- \frac{1}{V_{ijk}} (h_{RAD(x,y,z,t)} + h_{CONV(x,y,z,t)} + h_{ehap(x,y,z,t)}),
$$

$$(3.15)$$

where, $T_{(x,y,z,t)}$ is the instantaneous temperature of the tissue (°C) at the point x, y, z and at the time t. $C_{(x,y,z)}$ is the tissues located at (x, y, z), $k_{(x,y,z)}$ is the specific heat (J/kg°C), $hm_{(x,y,z)}$ is the thermal conductivity (W/m°C), and $B_{(x,y,z)}$ is the metabolism of the tissue (W/m³) and the blood perfusion parameter (W/m³ °C) Tb is the blood temperature. h_{RAD}, h_{CONV}, and h_{evap} are, respectively, the radiated, convective, and evaporation heat losses from peripheral cells (W), V is the increment of volume centered on x, y, z.

3.4 Channel Model

The fundamental step to build any wireless system is to study the propagation channel and model it precisely with respect to the environment and applications. The structure of the channel model has a strong influence on the system performance assessment. In case of channel modeling for WBAN, there are a number of measurement campaigns and path loss studies [55–63], which are mostly on UWB system, and some have also considered implant communication. However, in [64], channel modeling for wide range of frequencies has been discussed, varied from very low frequency of 13.5 MHz to 10.6 GHz in UWB band, including medical implant communications service (MICS) in the frequency range of 402–405 MHz. Most published works on channel modeling for WBAN communication are dealing with UWB frequency band due to short range and low power. On the other hand, the narrowband wireless propagation indicates that the channel under consideration is sufficiently narrow that its frequency response can be considered flat. In [65], transmission when the inverse of the signal bandwidth is much greater than the propagation path delays is termed narrowband.

In order to develop an accurate WBAN channel model, it is important to study the propagation mechanism of wireless radio waves on the body. Such a study will make known the fundamental propagation characteristics of WBAN. The WBAN signals usually consist of multipath components, arising from diffusion, diffraction, and scattering due to surrounding body and obstacles. Each multipath component can be thought as an independent traveling plane wave, whose amplitude, phase, incoming angle, and time delay are random variables. The WBAN channel model characterizes the path loss of WBAN devices taking into account possible shadowing due to the human body or obstacles near the human body and postures of human body.

3.4.1 Path Loss Model

Path loss is a fundamental characteristic of electromagnetic wave propagation and is utilized in system design, in order to predict system coverage. Usually, path loss is examined using the Friis transmission formula [19] which provides a means for predicting the received power. The path loss model in (*dB*) between the transmitting and the receiving antennas as a function of the distance (*d*) in free space [66, 67] is described by:

$$PL(d) = PL_0 + 10n \log_{10}\left(\frac{d}{d_0}\right), \qquad (3.16)$$

where PL_0 is the path loss at a reference distance d_0 and n is the path loss exponent.

The path loss near the antenna depends on the separation between the antenna and the body due to antenna mismatch. Unlike traditional wireless communications, the path loss for WBAN is both distance and frequency dependent. The frequency dependence is an antenna effect.

3.4.2 On-Body Measurement and Modeling

The propagation characteristics of the WBAN channel were measured for on-body to off-body and on-body to on-body, using a vector network analyzer (VNA) [68, 69]. The complex impulse response was calculated from measured complex transfer function in the frequency domain by fast Fourier transform

(FFT) function of the instrument in the frequency range of 3.1–10.6 GHz for UWB band. Measurements were done in an office environment for medical healthcare applications and in a hospital room.

3.4.2.1 Office room

The office room is having metal walls, windows, and its floor made of concrete board which is covered by carpet. The office room is furnished by desks, chairs, and PC monitors as shown in Figure 3.22.

The measurement configuration is shown in Figure 3.23 with positions of transmitting (Tx) and receiving (Rx) antennas. Tx position was fixed near the wall, and Rx positions were changed to the different locations. To have a better understanding of shadowing by human body, the human body (Rx antennas) is rotated to the different angles with respect to the fixed Tx antenna. The Rx antenna is placed on the front of body near to umbilicus, as shown in Figure 3.24.

Figure 3.25 shows the set up of antenna measurements to evaluate the effect of body on antenna patterns. To have an accurate model, antenna patterns in free space and on the surface of human body are compared in Figure 3.26.

Antenna has omni-directional pattern in horizontal plane for free space as shown in Figure 3.26(a). However, on-body patterns of antenna with different distance from body, which are 1, 2, 3, and 4 cm, show that the front side is almost same as free space as shown in Figure 3.26(b). But, the antenna gain in 180 degree is decreased around 20 dB, which is the effect of body shadowing. Therefore, on the time of measurements antenna patterns were only changed within ±60 degree from the center of back region.

On-body antenna characteristics were measured in an anechoic chamber, while channel measurements were done in office environment. The Tx antenna

Fig. 3.22 The office room for measurements.

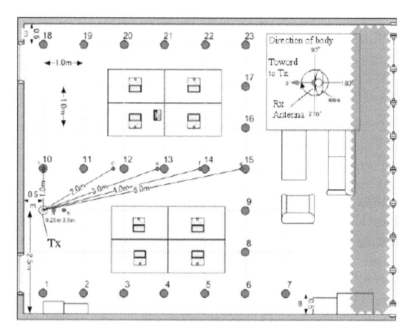

Fig. 3.23 Office room measurements configuration.

Fig. 3.24 On-body (Rx) antenna and its location on body surface.

is fixed near to wall, while the Rx antenna (on-body) positions were changed in office area. The effect of ground is considered in measurements. All data were averaged for statistical analysis. Since transmission distance is different

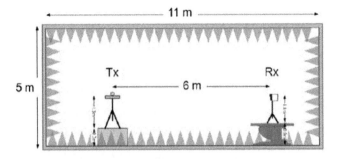

Fig. 3.25 Antenna measurement set up.

for each Rx position, all responses were normalized by the maximum value, and the delay time of the first response was shifted to 0.

The proposed model is as follows:

$$h(t) = \sum_{m=0}^{L-1} \alpha_m \delta(t - \tau_m), \qquad (3.17)$$

$$|\alpha_m|^2 = \Omega_0 e^{-\frac{\tau_m}{\Gamma} - k[1-\delta(m)]} \beta, \qquad (3.18)$$

$$k = \Delta k (\ln 10/10), \qquad (3.19)$$

$$\tau_0 = d/c, \qquad (3.20)$$

$$\beta \sim \log \, normal(0, \sigma), \qquad (3.21)$$

where $h(t)$ is the complex impulse response and L is number of arrival paths, which is modeled as Poisson random variable with mean $\bar{L} = 400$. a_m is the amplitude of each path, τ_m with $m = 1 .... L - 1$ is timing of path arrivals and modeled as Poisson random process with arrival rate $\lambda = 1/(0.50125 \, ns)$. k is the effect of K-factor for non-line-of-sight (NLOS) and Ω_0 is the path loss which depends on the environment and line-of-sight situation. Finally, d is the distance between Tx and Rx, and c is the velocity of light. Table 3.3 shows parameters for each body direction.

3.4.2.2 Hospital room

The layout of hospital room is shown in Figure 3.27 [70]. The room consists of four beds, cabinets, and medical equipments.

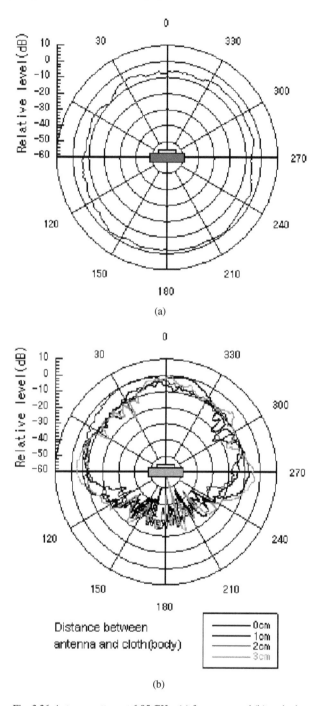

Fig. 3.26 Antenna pattern at 6.85 GHz: (a) free space and (b) on-body.

Table 3.3. Parameters for Each Direction of Body.

Direction of body	Γ (ns)	$k(\Delta k\ [dB])$	σ (dB)
0	44.6346	5.111 (22.2)	7.30
90	54.2868	4.348 (18.8)	7.08
180	53.4186	3.638 (15.8)	7.03
270	83.9635	3.983 (17.3)	7.19

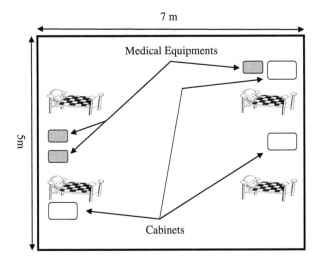

Fig. 3.27 Hospital room measurement set up.

The possible transceiver locations on the body are shown in Figure 3.28. It indicates that there are many paths in an on-body network communication to collect different vital signs. The locations on the body in the measurements are chosen to cover most of the applications shown in the IEEE.802.15.6 application matrices [71].

Figure 3.29(a)–(b) shows an example of the attachment preparations for measurements in the hospital room.

Distance between two on-body antennas is shown in Table 3.4. It can be seen that two sets of measurement, in an anechoic chamber and in a hospital room, have been performed for comparisons where there are no reflections from surrounding environment. Through all the measurements, 15 mm separation is kept between body surface and antenna. The height and weight of the volunteer participating in this experiment are 173 cm and 64 kg, respectively.

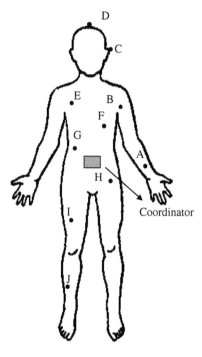

Fig. 3.28 Locations of antennas and coordinator on the body.

Table 3.4.   Distance (mm) between Two On-Body Antennas.

Position	Position index	Anechoic chamber	Hospital room
Left-hand side	A-1	430	410
Left-hand front	A-2	402	390
Left upper arm	B	360	424
Left ear	C	582	570
Head	D	722	742
Shoulder	E	390	402
Chest	F	236	258
Right lower chest	G	155	176
Left waist	H	184	180
Thigh	I	402	400
Ankle	J	986	984

Measured path loss when Rx antennas are attached at different locations is shown in Figures 3.30 and 3.31 for an anechoic chamber and a hospital room, respectively. In WBAN as antenna characteristics are highly effected by human body, the antenna gain at both Tx and Rx is not removed.

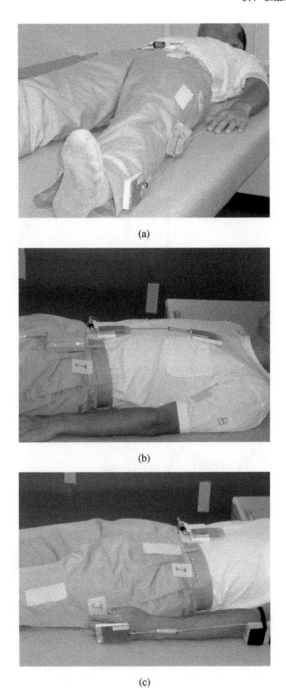

Fig. 3.29 Attachment of antennas and coordinator on the body.

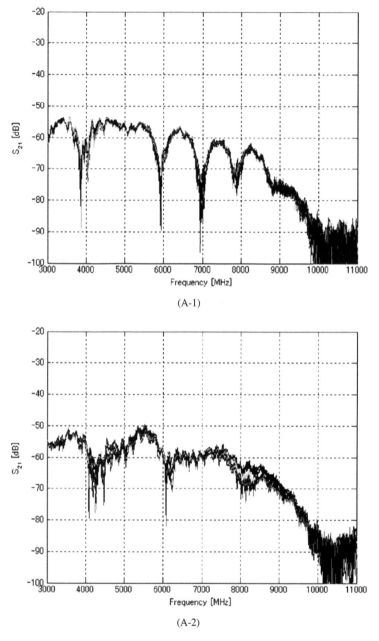

(A-1)

(A-2)

Fig. 3.30 (*Continued*)

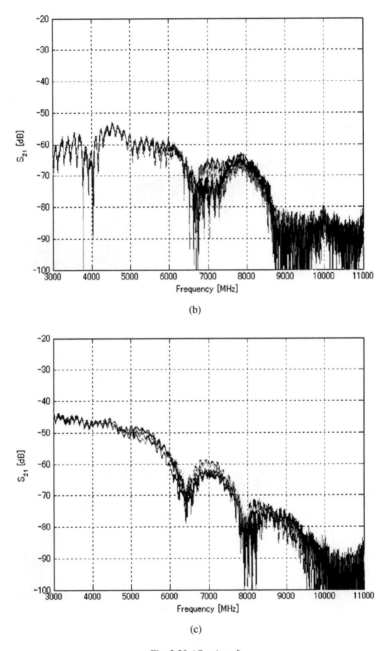

(b)

(c)

Fig. 3.30 (*Continued*)

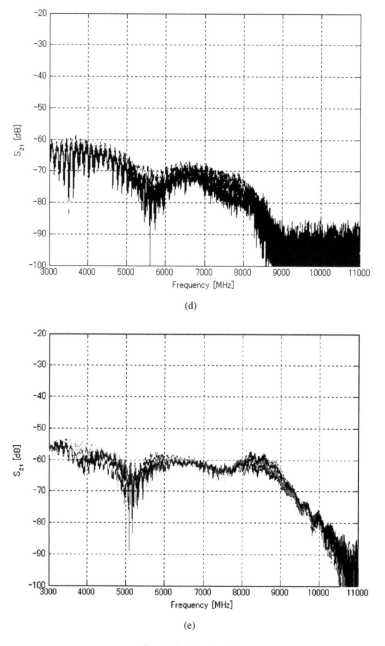

(d)

(e)

Fig. 3.30 (*Continued*)

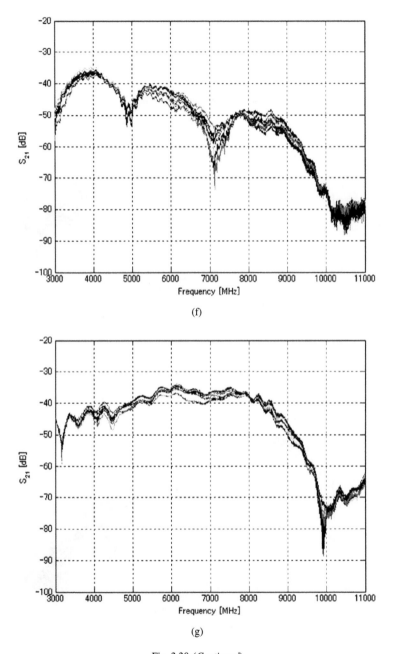

(f)

(g)

Fig. 3.30 (*Continued*)

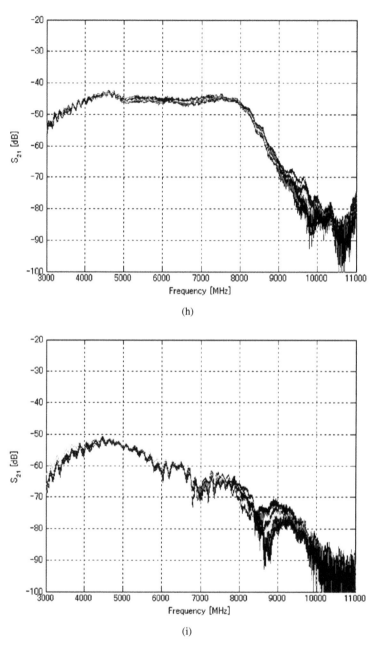

(h)

(i)

Fig. 3.30 (*Continued*)

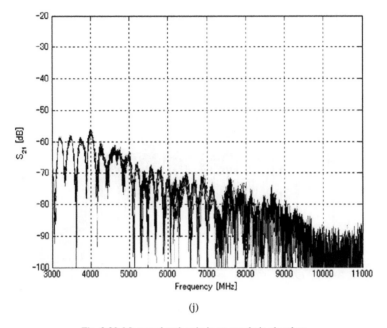

(j)

Fig. 3.30 Measured path gain in an anechoic chamber.

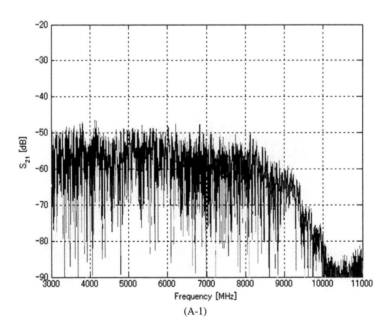

(A-1)

Fig. 3.31 (*Continued*)

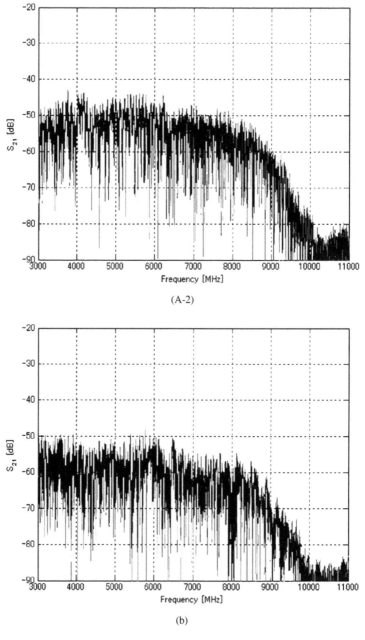

(A-2)

(b)

Fig. 3.31 (*Continued*)

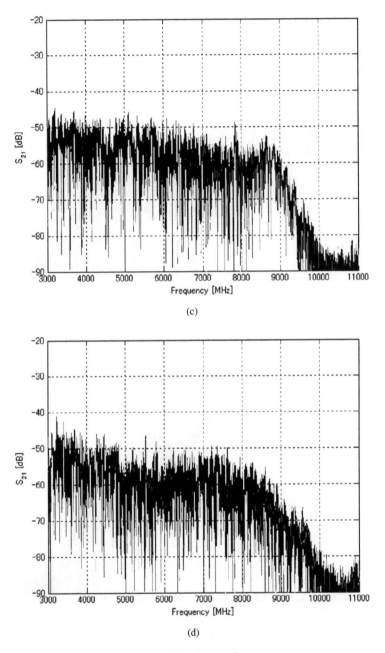

(c)

(d)

Fig. 3.31 (*Continued*)

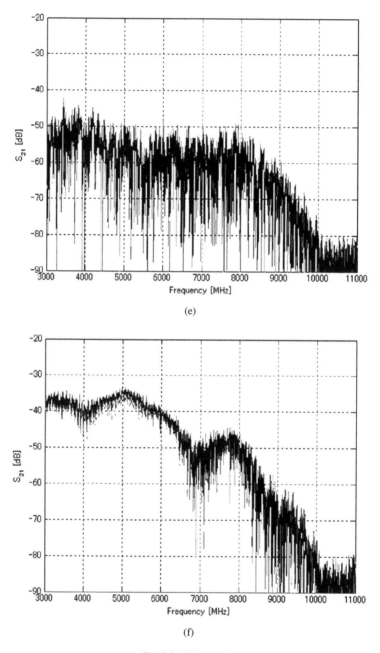

(e)

(f)

Fig. 3.31 (*Continued*)

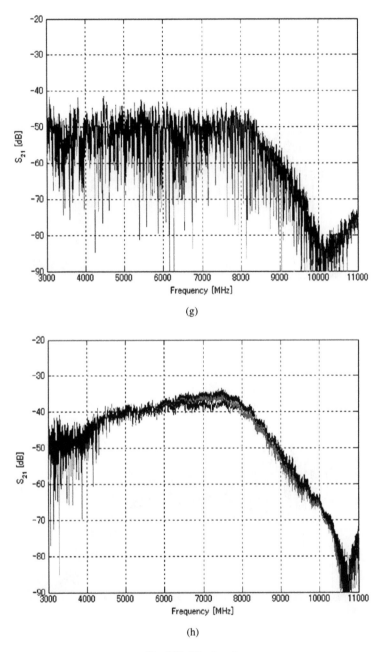

(g)

(h)

Fig. 3.31 (*Continued*)

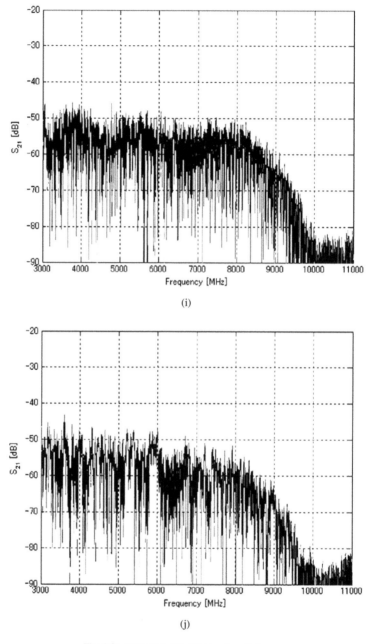

(i)

(j)

Fig. 3.31 Measured path gain in a hospital room.

The path gain at a frequency f, position p, and j-th snapshot is obtained by:

$$PL_j^p(f) = 10\log_{10}|H_j^p(f)|^2, \quad p \in \{A-1, A-2, \ldots, J\}, \quad (3.22)$$

where $H_j^p(f)$ is the measured S_{21} in [dB].

The path loss model simply can be obtained by:

$$L_{\text{path}}(d) = a \cdot \log_{10} d + b + N, \quad (3.23)$$

where $L_{\text{path}}(d)$ is the path loss in [dB] at the distance d (mm). N is a stochastic term which has a log-normal distribution with zero-mean and standard deviation of σ_N. a and b are parameters derived by a least-square fitting to the measured average path loss over the frequency range and given by $l_{\text{path}j}^p(d)$ [dB]:

$$l_{\text{path}j}^p(d) = -10 \cdot \log_{10}\left\{\frac{1}{N_F}\sum_{m=1}^{N_F} PL_j^p(f(m))\right\}, \quad (3.24)$$

where f_m stands for a frequency which corresponds to the m-th sample point at the measurements.

The path loss models in an anechoic chamber and in a hospital room are shown in Figure 3.32.

The results indicate that the path loss in the existence of the human body is often higher in the anechoic chamber. It is pointing out the strong influence of human body on RF signals regardless of a normal room or an anechoic chamber.

3.5 Summary of Chapter

In this chapter we have presented and discussed on antenna, electromagnetic and effect of human body, computational methods, specific absorption rate and thermal effect and finally channel model for WBAN.

In Section 3.2 antennas for WBAN were discussed. It was shown that magnetic antenna would be suitable for BAN applications, considering the long-term use of antenna on body surface. An omni-directional UWB antenna has been designed specifically for WBAN application. The antenna performance in the free space and close to the human body was presented. We have shown that the body absorbs a significant amount of the output power and changes the antenna characteristics, when an antenna is placed on the surface

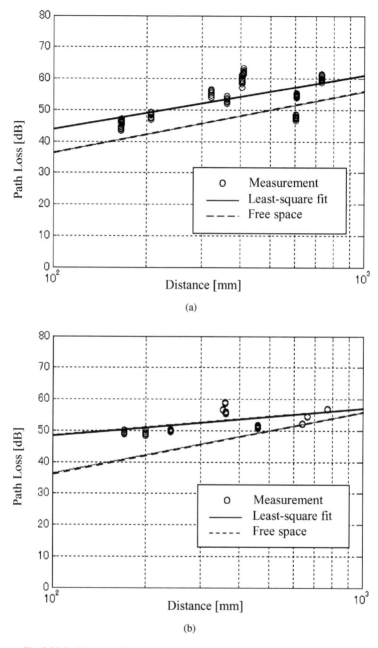

Fig. 3.32 Path loss model in an (a) anechoic chamber and in a (b) hospital room.

of human body. Therefore, destructively affects the performance of the system. Particularly, the human body affects the radiation pattern of antenna. The antenna characteristics will be changed, if antenna is designed in free space and used on body surface.

Section 3.3 presents detailed information on electromagnetic radiation and human body covering different tissues electrical properties at different frequency band. Computational methods such as MoM, FEM, and FDTD were described. Moreover, specific absorption rate and thermal effect have been discussed too.

The WBAN channel modeling and characterization were presented in Section 3.4. The model is based on extensive measurement results for a office room and a hospital room, and an anechoic chamber. The model includes the effect of antenna and user proximity.

The WBAN scenarios for on-body to off-body and on-body to on-body have been presented. In case of on-body to off-body, Tx position was fixed near the wall, and Rx positions were changed on the human body. In this measurement, human direction was also changed for consideration of shadowing by human body. For the case of on-body to on-body communications, the positions for the measurements are chosen in such way to cover most of the applications for WBAN. Path loss model and power delay profile have been extensively discussed for different positions.

References

[1] P. Salonen, Y. Rahmat-Samii, and M. Kivikoski, "Wearable antennas in the vicinity of human body," *IEEE Antennas and Propagation Society International Symposium*, vol. 1, pp. 467–470, June 2004.

[2] H. Adel, R. Wansch, and C. Schmidt, "Antennas for a body area network," *IEEE Antennas and Propagation Society International Symposium*, vol. 1, pp. 471–474, June 2003.

[3] B. Sanz-Izquierdo, F. Huang, and J. C. Batchelor, "Covert dual-band wearable button antenna," *Electronics Letters*, vol. 42, no. 12, pp. 668–670, June 2006.

[4] M. Klemm, I. Z. Kovcs, G. F. Pedersen, and G. Troester, "Novel small-size directional antenna for UWB WBAN/WPAN applications," *IEEE Transactions on Antennas and Propagation*, vol. 53, no. 12, pp. 3884–3896, December 2005.

[5] W. G. Scanlon, and N. E. Evans, "Numerical analysis of body worn UHF antenna systems," *IEE Electronics and Communications Engineering Journal*, vol. 13, no. 2, pp. 56–64, April 2001.

[6] M. Klemm, and G. Troester, "Textile UWB antenna for wireless body area networks," *IEEE Transactions on Antennas and Propagation*, vol. 54, no. 11, pp. 3192–3197, November 2006.

[7]   K. Yekeh Yazdandoost, and R. Kohno, "UWB antenna for wireless body area network," in *Asia Pacific Microwave Conference*, pp. 1647–1650, December 2006.

[8]   K. Yekeh Yazdandoost, and R. Kohno, "An antenna for medical implant communications system," in *European Microwave Conference*, pp. 968–971, October 2007.

[9]   K. Yekeh Yazdandoost, "A 2.4 GHz antenna for medical implanted communications," *Asia Pacific Microwave Conference*, pp. 1775–1778, December 2009.

[10] P. Gandhi, *Biological Effects and Medical Applications of Electromagnetic Energy*, Prentice Hall, NJ, 1990.

[11] S. Silver, *Microwave Antenna Theory and Design*, Peter Peregrinus Ltd., 1997.

[12] K. L. Wong, Y. G. Lee, and T. W. Chiou, "A low-profile planar monopole antenna for multiband operation of mobile handsets," *IEEE Transactions on Antennas and Propagation*, vol. 51, no. 1, pp. 121–125, January 2003.

[13] J. Perruisseau-Carrier, T. W. Hee, and P. S. Hall, "Dual-polarized broadband dipole," *IEEE Antenna and Wireless Propagation Letter*, vol. 2, pp. 310–312, 2003.

[14] K. L. Wong, *Compact and Broadband Microstrip Antennas*, Wiley, New York, NY, 2002.

[15] R. L. Li., E. M. Tentzeris, J. Laskar, V. F. Fusco, and R. Cahill, "Broadband loop antenna for DCS-1800/IMT-2000 mobile phone handsets," *IEEE Microwave and Wireless Components Letters*, vol. 12, pp. 305–707, August 2002.

[16] K. Yekeh Yazdandoost, and R. Kohno, "Ultra wideband L-loop antenna," in *IEEE International Conference on Ultra-Wideband*, pp. 201–205, September 2005.

[17] Z. N. Chen, and M. Y. W. Chia, "Broadband planar inverted-L antennas," *IEEE Proceedings on Microwaves, Antennas and Propagation*, vol. 148, pp. 339–342, October 2001.

[18] Z. N. Chen, and M. Y. W. Chia, "Suspended plate antenna with a pair of L-shaped strips," *IEEE Symposium on Antenna and Propagations*, vol. 3, pp. 64–67, June 2002.

[19] W. L. Stutzman, and G. A. Thiele, *Antenna Theory and Design*, Wiley, New York, NY, 1998.

[20] S. Yamamoto, T. Azakami, and K. Itakura, "Coupled nonuniform transmission line and its applications," *IEEE Transactions on Microwave Theory and Techniques*, vol. 15, pp. 220–231, April 1967.

[21] O. P. Rustogi, "Linearly tapered transmission line and its application in microwaves," *IEEE Transactions on Microwave Theory and Techniques*, vol. 17, pp. 166–168, March 1969.

[22] N. M. Martin and D. W. Griffin, "A tapered transmission line model for the feed-probe of a microstrip patch antenna," *IEEE Antenna and Propagations Symposium*, vol. 21, pp. 154–157, May 1983.

[23] I. Smith, "Principles of the design of lossless tapered transmission line transformers," in 7th *Pulsed Power Conference*, pp. 103–107, June 1989.

[24] Y. Wang, "New method for tapered transmission line design," *Electronics Letters*, vol. 27, pp. 2396–2398, December 1991.

[25] K. Murakami, and J. Ishii, "Time-domain analysis for reflection characteristics of tapered and stepped nonuniform transmission lines," in *Proceedings of the IEEE International Symposium on Circuits and Systems*, vol. 3, pp. 518–521, June 1998.

[26] K. Yekeh Yazdandoost, and R. Kohno, "Ultra Wideband Antenna," *IEEE Communications Magazine*, vol. 42, pp. S29–S32, June 2004.

[27] H. Schantz, *The Art and Science of Ultra Wideband Antennas*, Artech House, 2005.

[28] J. S. Colburn, and Y. Rahmat-Samii, "Electromagnetic scattering and radiation involving dielectric objects," *Journal of Electromagnetic Wave and Applications*, vol. 9, pp. 1249–1277, 1995.

[29] M. A. Jensen, and Y. Rahmat-Samii, "EM interaction of handset antenna and a human in personal communications," *Proceeding of IEEE*, vol. 83, pp. 7–17, January 1995.

[30] K .W. Kim, and Y. Rahmat-Samii, "Antennas and humans in personal communications: An engineering approach to the interaction evaluation," *Proceedings IEEE Engineering in Medicine and Biology Society*, pp. 2488–2491, October 1997.

[31] M. Okoniewski, and M. A. Stuchly, "A study of the handset and human body interaction," *IEEE Transactions on Microwave Theory and Technique*, vol. 44, pp. 1855–1864, October 1996.

[32] C. Gabriel, *The Dielectric Properties of Biological Materials, Radiofrequency Standards*, Plenum Press, New York, NY, 1994.

[33] C. H. Durney, H. Massoudi, and M. F. Iskander, *Radio Frequency Radiation Dosimetry Handbook*, Brooks Air Force Base, USAFSAM-TR- pp. 85–73, 1986.

[34] K. R. Foster, and H. P. Schwan, "Dielectric properties of tissues and biological materials: A critical review", *Critical Reviews in Biomedical Engineering*, vol. 17, pp. 25–104, 1989.

[35] D. M. Sullivan, "Frequency-dependent FDTD methods using Z transforms," *IEEE Transactions on Antennas Propagation.*, vol. 40, pp. 1223–1230, October 1992.

[36] A. Barnett, "The phase angle of normal human skin," *Journal of Physiology*, vol. 9, pp. 349–366, 1938.

[37] F. S. Branes, and B. Greenebaum, *Handbooks of Biological Effects of Electromagnetic Field*, Taylor and Francis, 2007.

[38] C. Gabriel, *Compilation of the Dielectric Properties of Body Tissues at RF and Microwave Frequencies*, Brooks Air Force Technical Report AL/OE-TR-1996-0037, 1996.

[39] Dielectric properties of body tissues, Institute for Applied Physics, Italian National Research Council, http://niremf.ifac.cnr.it .

[40] A. Bondeson, T. Rylander, and P. Ingelström, *Computational Electromagnetics*, Springer, 2005.

[41] B. S. Guru, and H. R. Hiziroglu, *Electromagnetic Field Theory Fundamentals,* Cambridge University Press, 2004.

[42] Pepper, D. W., J. C. Heinrich, *The Finite Element Method: Basic Concepts and Applications*, Taylor and Francis, Second ed., 2005.

[43] K. S. Yee, "Numerical Solution of Initial Boundary Value Problems Involving Maxwell's Equations in Isotropic Media," *IEEE Transactions on Antennas and Propagation*, vol. 14, pp. 302–307, May 1966.

[44] IEEE Standards Coordinating Committee 28 on Non-Ionizing Radiation Hazards: Standard for Safe Levels with Respect to Human Exposure to Radio Frequency Electromagnetic Fields, 3 KHz to 300 GHz, The Institute of Electrical and Electronics Engineers, New York, 1992.

[45] Y. Rahmat-Samii, and W. L. Stutzman (eds.), "Special issue on wireless communications," in *IEEE Transactions on Antennas and Propagation*, vol. 46, June 1998.

[46] IEEE Std 1528TM-2003, IEEE recommended practice for determining the peak spatial-average Specific Absorption Rate (SAR) in the human head from wireless communications devices: Measurement techniques, IEEE, New York, 2003.

[47] E. R. Adair, and R. C. Peterson, "Biological effects of radiofrequency/microwave radiation," *IEEE Transactions on Microwave Theory and Techniques*, vol. 50, pp. 953–961, March 2002.

[48] F. S. Barnes, and B. Greenebaum, *Bioengineering and Biophysical Aspects of Electromagnetic Fields*, Taylor and Francis, Third ed., 2006.

[49] Q. Tang, N. Tummala, S. Kumar, S. Gupta, and L. Schwiebert, "Communication scheduling to minimize thermal effects of implanted biosensor networks in homogeneous tissue," *IEEE Transactions on Biomedical Engineering*, vol. 52, pp. 1285–1294, July 2005.

[50] A. Hirata, M. Morita, and T. Shiozawa, "Temperature increase in the human head due to a dipole antenna at microwave frequencies," *IEEE Transactions on Electromagnetic Compatibility*, vol. 45, pp. 109–116, February 2003.

[51] O. P. Gandhi, Q.-X. Li, and G. Kang, "Temperature rise for human head for cellular telephones and for peak SARs prescribed in safety guidelines," *IEEE Transactions on Microwave Theory and Techniques*, vol. 49, pp. 1607–1613, September 2001.

[52] P. Bernardi, M. Cavagnaro, S. Pisa, and E. Piuzzi, "SAR distribution and temperature increase in an anatomical model of the human eye exposed to the field radiated by the user antenna in a wireless LAN," *IEEE Transactions on Microwave Theory and Techniques*, vol. 46, pp. 2074–2082, December 1998.

[53] P. Bernardi, M. Cavagnaro, S. Pisa, and E. Piuzzi, "Specific absorption rate and temperature increases in the head of a cellular phone user," *IEEE Transactions on Microwave Theory and Techniques.*, vol. 48, pp. 1118–1126, July 2000.

[54] A. Ibrahiem, and C. Dale, "Analysis of the temperature increase linked to the power induced by RF source," *Progress in Electromagnetics Research*, PIER 52, pp. 23–46, 2005.

[55] K. Takizawa, T. Aoyagi, J. Takada, N. Katayama, K. Yekeh Yazdandoost, T. Kobayashi, and R. Kohno, "Channel model for wireless for body area networks," *IEEE Conference on Engineering in Medicine and Biology Society*, pp. 1549–1552, August 2008.

[56] K. Sayrafian-Pour, W.-B. Yang, J. Hagedorn, J. Terrill, K. Yekeh Yazdandoost: "A statistical path loss model for medical implant communication channels," in *IEEE Conference on Personal, Indoor and Mobile Radio Communications Symposium*, September 2009.

[57] K. Sayrafian-Pour, K. Yekeh Yazdandoost, W.-B. Yang, J. Hagedorn, J. Terrill, "Simulation study of body surface RF propagation for UWB wearable medical sensors," *International Symposium on Applied Sciences in Biomedical and Communication Technologies*, pp. 1–6, November 2009.

[58] T. Zasowski, F. Althaus, M. Stager, A. Wittneben, and G. Troster, "UWB for noninvasive wireless body area networks: Channel measurements and results," in *Proceedings of the IEEE Conference on Ultra Wideband Systems and Technologies*, pp. 285–289, November 2003.

[59] A. Fort, J. Ryckaert, C. Desset, P.D. Doncker, P. Wambacq, and L.V. Biesen, "Ultrawideband channel model for communication around the human body," *IEEE Journal on Selected Areas in Communications*, vol. 24, pp. 927–933, April 2006.

[60] A. Sani, A. Alomainy, and Y. Hao, "Effect of the indoor environment on the UWB on-body radio propagation channel." in *European Conference on Antenna and Propagation*, pp. 455–458, March 2009.

[61] Y. Zhao, Y. Hao, A. Alomainy, and C. Parini, "UWB on-body radio channel modeling using ray theory and sub-band FDTD method," *IEEE Transactions on Microwave Theory and Techniques, Special Issue on Ultra-Wideband,* vol. 54, pp. 1827–1835, April 2006.

[62] J. Ryckaert, P. D. Doncker, R. Meys, A. D. L. Hoye, and S. Donnay, "Channel model for wireless communication around human body," *Electronic Letters,* vol. 40, 2004.

[63] A. Alomainy, Y. Hao, X. Hu, C. G. Parini, and P.S. Hall, "UWB on-body radio propagation and system modeling for wireless body-centric networks," *IEE Proceedings on Communications,* vol. 153, pp. 107–114, 2006.

[64] K. Yekeh Yazdandoost, and K. Sayrafian-Pour, "Channel model for body area network, IEEE P802.15-08-0780-09-0006," *IEEE Standards Association, IEEE 802.15 Working Group for WPAN,* https://mentor.ieee.org/802.15/documents, April 2009.

[65] A. G. Siamarou, "Digital transmission over millimeter-wave radio channels: A Review," *IEEE Antennas and Propagation Magazine,* vol. 51, pp. 196–203, December 2009.

[66] E. Reusens, W. Joseph, G. Vermeeren, and L. Martens, "On-body measurements and characterization of wireless communication channel for arm and torso of human," in *International Workshop on Wearable and Implantable Body Sensor Networks,* pp. 26–28, March 2007.

[67] A. Fort, J. Ryckaert, C. Desset, P. De Doncker, P. Wambacq, and L. Van Biesen, "Ultra-wideband channel model for communication around the human body," *IEEE Journal on Selected Areas in Communications,* vol. 24, pp. 927–933, April 2006.

[68] H. Sawada, T. Aoyagi, J. Takada, K. Yekeh Yazdandoost, and R. Kohno, "Channel model between body surface and wireless access point for UWB band," *IEEE 802.15-08-0576-00-0006,* August 2008.

[69] A. Takahiro, T., J. Takada, K. Takizawa, N. Katayama, T. Kobayashi, K. Yekeh Yazdandoost, H. Li, and R. Kohno, "Channel model for wearable and implantable WBANs," *IEEE 802.15-08-0416-04-0006,* November 2008.

[70] A. Takahiro, J. Takada, K. Takizawa, H. Sawada, N. Katayama, K. Yekeh Yazdandoost, T. Kobayashi, H. Li, and R. Kohno, "Channel model for WBANs," *IEEE 802.15-08-0416-03-0006,* September 2008.

[71] D. Lewis, "802.15.6 call for applications — response summary," *IEEE 802.15-08-0407-06-0006,* January 2009.

4

Physical Layer Technologies

The main stream in developing wireless communication networks is in two directions. One is the capability to support high-rate transmission and another is the capability to support mobility. When wireless body area network (WBAN) is used for medical or healthcare monitoring, or dealing with control signals in consumer electronics, the required data rate ranges from several 10 kbps to several hundreds kbps. However, the required date rate may increase to several Mbps and even more than 10 Mbps if video signal or high-definition television (HDTV) signal transmission is involved. This may happen when a capsule endoscope inside body sends captured video or HDTV pictures to a controller or a monitor outside the body in real time. It may even become an ordinary requirement when HDTV streaming is conducted among devices within a WBAN for consumer electronics. Technologies to increase channel capacity, such as multi-input multi-output (MIMO) used in wireless metropolitan area network (WMAN) and wireless local area network (WLAN), are not suitable to WBAN. Moreover, data rate beyond 10 Mbps is challenging for WBAN with limited frequency band. As WBAN devices are usually attached to or in peripheral of human body, WBAN channels are different from traditional ones. As will be shown in Chapter 5, WBAN can experience serious shadowing or fading. This increases the difficulty for designing WBAN.

In the open system interconnection (OSI) seven-layer's model defined by International Organization for Standardization (ISO), physical layer is the first layer, in which the source data is converted into electrical signal for transmission. In order to increase the power efficiency, bandwidth efficiency, as well as reliability for transmission, a series of processing, including channel coding and decoding, modulation and demodulation, filtering, and frequency conversion, are carried out. Mechanic specifications of interface to enable physical connections are also defined in this layer.

From the system design point of view, power efficiency, bandwidth efficiency, and system performance (including transmission data rate, communication range, error probability) are of the top importance. All these parameters are dependent on the electrical characteristics of physical layer (PHY) and will be the main focus in this chapter. Because the technologies for narrow band and those for UWB have different characteristics, narrow band and UWB will be discussed in separate sections. Some developed standards or specifications for short-range radio will be overviewed in the sense of their suitability for WBAN.

4.1 General Concepts

We start with a short introduction on the concept of WBAN from an upper layer point of view. Then, some basic concepts surrounding digital communication systems are addressed.

4.1.1 Concept of WBAN

The intent of WBAN is to provide short-range wireless links among devices inside, on, or in proximity of body [1]. Generally, a WBAN is composed of a coordinator and several nodes as shown in Figure 4.1. The coordinator is responsible for network forming, medium access control (MAC), as well as channel assignment. The details are described in Chapter 5. Due to the short communication range, the basic network topology is a one-hop star. In which, the coordinator is located at the topological center and all other nodes communicate with coordinator directly. However, in some cases, extended two-hop connection through tree topology may be required to combat shadowing or fading. Although it may be rare, three-hop should also be available in general. In a tree topology, some nodes in the network must act as relay nodes.

From PHY point of view, there can be two types of WBAN devices. One is full function device (FFD) and another is reduced function device (RFD). An FFD can take the tasks such as network control and channel assignment and can communicate with several other devices. Only an FFD can play the role of a coordinator or a relay node. In contrast, an RFD only communicates

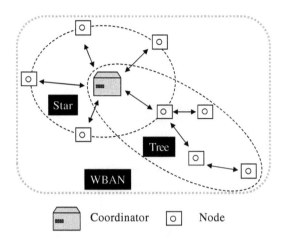

Fig. 4.1 Conceptual sketch of a WBAN.

with a neighbor device and can only be used as an end node. All devices can transmit and receive, thus provide bi-directional communications.

Depending on their operating locations with body, WBAN can be categorized as wearable BAN and implant BAN. A wearable BAN mainly operates on the surface or in peripheral proximity of body. Wearable BAN has a root from a so-called wearable network [2], which put more emphasis on computing aspects. In contrast, an implant BAN operates inside body. Implant BAN is originated from medical trials [3]. Wearable BAN and implant BAN have some different aspects. First, wearable BAN experiences multipath channel including shadowing while implant BAN mainly suffers from significant decay in propagation because of the effects of body tissues. Second, implant BAN is more power restricted and requires smaller form factor compared to wearable BAN. Moreover, implant BAN can use MICS band, which is basically only assigned to medical implant communications. However, it is desirable to enable interoperability between wearable BAN and implant BAN when they are used for same medical or health purposes.

4.1.2 General Transceiver Structure

The basic structure of a PHY in digital communication systems includes a transmitter and a receiver. Both are implemented together in a single device (a coordinator or a node) to enable bi-directional communications.

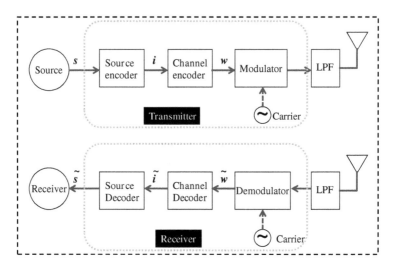

Fig. 4.2 General transceiver model.

The main task of transceivers is to transmit and receive data between a pair of devices with high reliability and high efficiency. A general model of a digital transceiver is shown in Figure 4.2.

4.1.2.1   Transmitter

The main operations in the transmitter are source encoding, channel encoding, and modulation. With the source encoding, source data is efficiently converted into digital representation. Generally, binary digits 0 and 1 are employed using either natural binary code or Gray code. Suppose the length of a code is N_b, the natural binary code can represent a decimal digit up to $2^{N_b} - 1$. The natural binary code can be transformed to Gray code using:

$$g_k = \begin{cases} b_{k+1} \oplus b_k; & k = 1, 2, \ldots, N_b - 1 \\ b_k; & k + N_b \end{cases}, \tag{4.1}$$

where b_k and g_k denote the bits of the natural binary code and Gray code, respectively. Two neighbor bits in a Gray code have only one different bit, which is favorable in many applications. The source data usually contains redundancy. One main objective of source encoding is to reduce source redundancy. The source data after source encoding can be denoted as bit block with length of k, i.e., $i = (i_1, i_2, \ldots, i_k)$.

The purpose of channel coding is to add "controlled redundancy" to the data bit block i to combat signal distortion and fading occurred during transmission in channel. The channel coding is usually referred to as forward error correction (FEC). If $n - k(n > k)$ redundant bits are added to the data bit block of $i = (i_1, i_2, \ldots, i_k)$ to generate a codeword $w = (w_1, w_2, \ldots, w_n)$, the length of a codeword is n and k/n is defined as coding rate. The larger the coding rate, the higher the efficiency of an FEC code.

Modulation is a conversion to convey baseband data to a higher frequency band for efficient transmission. In the most basic forms, the baseband data bits are used to modulate one of the three factors of a carrier, i.e., amplitude, frequency, or phase, to obtain a modulated signal waveform. The produced corresponding modulations are called as amplitude shift keying (ASK), frequency shift keying (FSK), and phase shift keying (PSK). Combinations between some of these three basic modulations are available to provide enhanced performance. Examples of the combinations include amplitude-phase shift keying (APSK), amplitude-frequency shift keying (AFSK), quadrature amplitude modulation (QAM), etc.

4.1.2.2 Receiver

The task of receiver is to recover the data bits sent by the transmitter from the received signal waveform. There are basically two types of operations in the demodulator. One is coherent detection and another is non-coherent detection. Coherent detection requires an exact carrier reference with the same frequency and phase as those of the transmitter. In contrast, there is no need of exact frequency or phase information in non-coherent detection. Therefore, non-coherent receiver is much less complex and low cost than coherent receiver. However, coherent receiver requires less signal-to-noise ratio (SNR) than non-coherent detection to achieve the same error probability. Generally, E_b/N_0 is used as a measure of SNR, where E_b is the energy per data bit and N_0 is the power spectral density of noise. The expression of error probability is bit error rate (BER). The required E_b/N_0 to achieve a given BER is fixed if a modulation scheme is decided. The required BER is application dependent. For voice, a BER of $10^{-2} - 10^{-3}$ is acceptable. For video, the BER should be smaller than 10^{-6}. For data, the BER is usually smaller than 10^{-9}. Channel coding must be involved to achieve such small BERs.

The function of channel decoder is to delete the "controlled redundancy" added by the channel encoder and recover the data bits block of $\tilde{i} = (\tilde{i}_1, \tilde{i}_2, \ldots, \tilde{i}_k)$ from the demodulated codeword $\tilde{w} = (\tilde{w}_1, \tilde{w}_2, \ldots, \tilde{w}_n)$.

Finally, the source decoder recovers the source data from $\tilde{i} = (\tilde{i}_1, \tilde{i}_2, \ldots, \tilde{i}_k)$.

4.1.2.3   Filter

The low pass filters (LPFs) shown in Figure 4.2 are used for signal spectrum shaping and to cut out-band components. An ideal LPF should have constant amplitude within the desired bandwidth and cut frequency components out of the bandwidth. Moreover, the phase should show linearity within the desired bandwidth. The most popular LPF used is the square-root raised cosine filter (SRCF). The product of the two SRCFs at the transmitter and receiver produces a frequency transfer function given by:

$$H(f) = \begin{cases} T; & 0 < |f| \le \dfrac{1-\alpha}{2T} \\ \dfrac{T}{2}\left\{1 - \sin\left[\dfrac{\pi T}{\alpha}\left(f - \dfrac{1}{2T}\right)\right]\right\}; & \dfrac{1-\alpha}{2T} < |f| \le \dfrac{1+\alpha}{2T} \\ 0; & \text{others} \end{cases}$$

(4.2)

The corresponding time-domain impulse response is raised cosine function of:

$$h(t) = \frac{\sin\left(\frac{\pi t}{T}\right)}{\frac{\pi t}{T}} \cdot \frac{\sin\left(\frac{\pi \alpha t}{T}\right)}{1 - \left(\frac{2\alpha t}{T}\right)^2},$$

(4.3)

where α is called as roll-off factor and $0 \le \alpha \le 1$.

Frequency transfer function and time-domain impulse response of raised cosine filter (the product of the transmitter and receiver SRCFs) are shown in Figure 4.3. When $\alpha = 0$, we get the ideal performance of the pair of filters.

4.1.3   Link Budget

Link budget is the calculation of power level evolution of whole transmission chain, i.e., from transmitter to receiver. Link budget is calculated when designing a communication system. The purpose of link budget calculation is to decide reasonable system parameters by considering trade-off among performance, complexity, and cost, while making sure that the transmitted signal

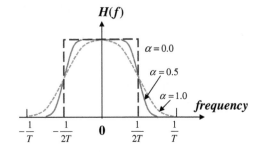

(a) Frequency transfer function

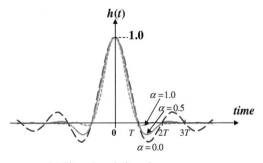

(b) Time-domain impulse response

Fig. 4.3 Raised cosine frequency transfer function and time-domain impulse response.

from transmitter can be correctly received at receiver. Performance parameters specified from an application are generally BER and data rate. When BER and modulation are decided, the required E_b/N_0 is fixed as discussed in Section 4.1.2.2. Usually, the designed E_b/N_0 is set to a much larger value than the required value to provide a reasonable link margin in accordance to operating environment.

Other system parameters used in link budget calculation are given as follows:

- Average transmit power (P_T)
- Transmit antenna gain (G_T)
- Propagation loss or path loss (L_F)

  When the propagation is over free space, the path loss is calculated using:

$$L_F = 20\log\left(\frac{4\pi d}{\lambda}\right) \ (dB), \qquad (4.4)$$

where λ is the wavelength at the carrier frequency and d the distance between transmitter and receiver.

- Receive antenna gain (G_R)
- Receiver noise figure (N_f)

which is a parameter to describe the electrical characteristics of receiver circuit defined by:

$$N_f = \frac{S_{in}/N_{in}}{S_{out}/N_{out}}. \tag{4.5}$$

- Receiver noise power spectral density (N_0)

which is a parameter related to equivalent input noise temperature. It is calculated as:

$$N_0 = k + 10\log(T_i) \ (dB/Hz), \tag{4.6}$$

where T_i is the absolute temperature in Kelvin (K) and k is Boltzmann's constant given by:

$$k = 10\log(1.38 \times 10^{-23}) = -228.6 \ (dB/Hz). \tag{4.7}$$

T_i has a value of 290 K in room temperature, which results in

$$N_0 = -204 \ \text{dBW or} \ N_0 = -174 \ \text{dBm} \tag{4.8}$$

- Implementation loss (I)

With the above parameters, the received power is calculated as:

$$P_r(dBm) = P_t(dBm) + G_t(dBi) + G_r(dBi) - L_f(dB). \tag{4.9}$$

Then C/N_0 provided by the link is:

$$C/N_0(dB \cdot Hz) = P_r(dBm) - N(dBm) - N_f(dB) - I(dB), \tag{4.10}$$

which can be changed to E_b/N_0 using:

$$E_b/N_0 = C/N_0 - 10\log(R_b), \tag{4.11}$$

where R_b denotes the data rate. Comparing the result of Equation (4.11) with the required E_b/N_0, the link margin can be obtained. The larger the link margin, the more reliable the communication link.

4.2 Narrow Band Modulations

Many traditional narrow band modulation technologies [4–7] have been developed and widely used in various wireless communication networks. WBAN transceivers generally operate on a button battery or even further small-sized batteries. Therefore, they are much more power restricted. From the power efficiency stand, modulations with small envelope variation are favorable. In this section, several power-efficient modulations are introduced.

4.2.1 Phase Modulations

4.2.1.1 PSK and DPSK

In PSK modulation, data bits are used to modulate the phase of a carrier. A binary PSK (BPSK) signal can be denoted using the following expression:

$$S_{BPSK}(t) = \begin{cases} A\cos(2\pi f_c t); & data = 1 \\ A\cos(2\pi f_c t + \pi); & data = 0, \end{cases} \quad (4.12)$$

where phases 0 and π are employed in the modulation. By increasing the number of phases, M-ary PSK modulations can be obtained as shown in Figure 4.4.

PSK signal can be detected using coherent demodulation or differentially coherent demodulation. The latter is usually referred to as differential PSK (DPSK). In coherent demodulation exact carrier is required, which is difficult or needs complex extracting circuit such as phase-locked loop when channel undergoes rapid variation. In DPSK, suppose that the current symbol is $A\cos(2\pi f_c t + \varphi_k)$, the forward neighbor symbol $A\cos[2\pi f_c(t - T) + \varphi_{k-1}]$ is used as the demodulating reference signal. Then, only the phase difference between two neighbor symbols is needed in demodulation. The complexity of the receiver can be greatly reduced compared to that of coherent demodulation.

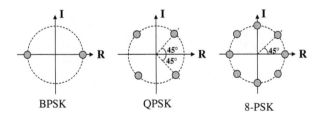

Fig. 4.4 Signal constellation of M-ary PSK.

The error probability of BPSK and differential BPSK (DBPSK) is given as follows [5]:

$$P_{\text{BPSK}} = \frac{1}{2} erfc \left(\sqrt{\frac{E_b}{N_0}} \right) \quad \text{(without differential encoding),} \quad (4.13)$$

$$P_{\text{BPSK}} = erfc \left(\sqrt{\frac{E_b}{N_0}} \right) \left(1 - \frac{1}{2} erfc \left(\sqrt{\frac{E_b}{N_0}} \right) \right)$$
$$\text{(with differential encoding),} \quad (4.14)$$

$$P_{\text{DBPSK}} = \frac{1}{2} e^{-(E_b/N_0)},$$

where $erfc(x)$ is the complementary error function defined by:

$$erfc(x) = \frac{2}{\sqrt{\pi}} \int_x^\infty \exp(-u^2) \, du. \quad (4.15)$$

The error probability of DBPSK is larger than that of BPSK. A multiple-symbol differential detection can be used to improve the error performance of conventional DPSK [8].

4.2.1.2 $\pi/4$-Shift QPSK

To reduce the envelope variation of quadrature PSK (QPSK) modulation, $\pi/4$-shift QPSK can be employed. By adding a $\pi/4$-radians phase rotator to a QPSK modulator, $\pi/4$-shift QPSK can be obtained. The signal constellation and phase-shifting diagram of a $\pi/4$-shift QPSK and a traditional QPSK are shown in Figure 4.5 together for comparison.

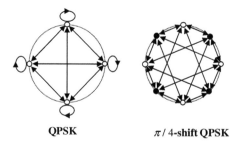

QPSK $\pi/4$-**shift QPSK**

Fig. 4.5 Signal constellation and phase-shifting diagram.

It can be seen that the phase shifts between neighbor symbols of QPSK are $0, \pm\frac{\pi}{2}$, and π, respectively. When phase shift is π, the signal envelope passes through the origin (0). In contrast, $\pi/4$-shift QPSK alternately uses two signal sets. One is $(0, \pm\frac{\pi}{2}, \pi)$ given in white circles and another is $(\pm\frac{\pi}{4}, \pm\frac{3\pi}{4})$ given in black circles in Figure 4.5. The second signal set is $\pi/4$-shift version of the first signal set. Therefore, the phase shift between neighbor symbols of $\pi/4$-shift QPSK does not pass through the origin and results in smaller envelope variation than QPSK. Another advantage of $\pi/4$-shift QPSK is that at least $\pi/4$ of phase shift occurs even the same symbols are sent continuously. This characteristic may facilitate the symbol-timing recovery.

4.2.1.3 Offset QPSK

Offset QPSK is another phase modulation method to reduce the envelope variation of QPSK. In QPSK, the in-phase bit (I-bit) and quadrature-phase bit (Q-bit) change at the same time. In OQPSK, the Q-bit is delayed in time by 1/2 bit interval from the I-bit or vice versa. Therefore, I-bit and Q-bit will not change at a same time. As a result, the phase shift between two neighbor QPSK symbols can only be 0 or $\pm\pi/2$. The phase shift diagram of OQPSK is shown in Figure 4.6. It can be seen that the restricted phase shift produces small peak-to-average ratio of envelope, i.e., small envelope variation.

4.2.2 Frequency Modulations

4.2.2.1 FSK and MSK

Frequency shift keying (FSK) is favorable in the sense of providing constant envelope. FSK uses two different carriers to represent data 0 and 1. Suppose

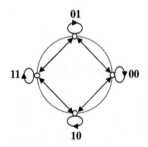

Fig. 4.6 Phase-shifting diagram of OQPSK.

the difference between the two carriers is Δf, FSK can be denoted by:

$$\begin{cases} s(t)|_{\text{data}=1} = \cos(2\pi f_c + \pi \Delta f)t \\ s(t)|_{\text{data}=0} = \cos(2\pi f_c - \pi \Delta f)t \end{cases}. \tag{4.16}$$

In order to perform optimum detection, the two signals in Equation (4.16) must be in vertical, i.e.,

$$\int_0^T \cos[(2\pi \Delta f_c + \pi \Delta f)t] \cdot \cos[(2\pi f_c - \pi \Delta f)t]\mathrm{d}t = 0. \tag{4.17}$$

The left-hand side of Equation (4.17) can be further calculated as:

$$\frac{1}{2}\int_0^T [\cos(4\pi f_c t) + \cos(2\pi \Delta f t)]\mathrm{d}t \cong \frac{1}{2}\frac{\sin(2\pi \Delta f T)}{2\pi \Delta f T}; \quad \text{for } f_c \gg T. \tag{4.18}$$

To let Equation (4.18) be equal to 0, the equation $2\pi \Delta f T = n \cdot \pi$ must be satisfied. Defining $h = \Delta f T$ as modulation coefficient, we can obtain a number of FSK modulations with different value h. The minimum h that satisfies the condition is 0.5. We call the modulation with $h = 0.5$ as minimum shift keying (MSK). By modifying Equation (4.16), an MSK signal can be denoted as:

$$S_{\text{MSK}}(t) = \cos[2\pi f_c t + \theta(t)]; \quad iT \leq t \leq (i+1)t, \tag{4.19}$$

and $\theta(t)$ is given by:

$$\theta(t) = \frac{\pi a_i}{2T}(t - iT) + \varphi_i = \frac{\pi a_i}{2T}t + \varphi_i', \tag{4.20}$$

where φ_i' only takes values of 0 or π.

Besides the constant envelope, side lobes of the power spectrum of MSK decay much more rapidly than that of PSK modulation. As examples, the power spectrums of MSK and QPSK are show in Figure 4.7. It can be seen that MSK presents less out-band emissions.

4.2.2.2 Gaussian Filtered FSK and MSK

One disadvantage of FSK or MSK is the wide bandwidth occupancy, which shows a wider main lobe than QPSK as can be seen in Figure 4.7. In order to increase the bandwidth efficiency, a Gaussian-type filter is usually used to shape the baseband rectangular signal. After smoothed by a Gaussian filter, the

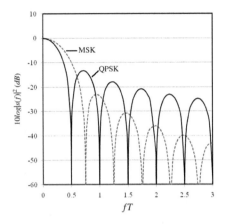

Fig. 4.7 Power spectrums comparison.

instant large frequency components can be greatly suppressed and bandwidth occupancy can be narrowed. Applying a Gaussian filter to FSK or MSK, we can get a Gaussian filtered FSK (GFSK) or a Gaussian-filtered MSK (GMSK).

The frequency transfer function of a Gaussian filter can be written as:

$$G(f) = \exp\left[-2\ln 2\left(\frac{fT}{BT}\right)\right],\qquad(4.21)$$

where B denotes the 3-dB bandwidth. The smaller the value of BT, the narrower the signal bandwidth. By providing both constant envelope and narrow bandwidth, GFSK and GMSK are favored when both power efficiency and frequency efficiency are required. However, it should be noted that reducing the value of BT may introduce intercode interference. Trade-off must be made in accordance to the requirement raised from application.

4.2.3 Spread Spectrum

Spread spectrum is a well-known technology for its capability of high time resolution, robustness against interference, and providing secure communication. Spread spectrum techniques include direct sequence spread spectrum (DSSS) and frequency hopping spread spectrum (FHSS). DSSS is more popular compared to FHSS. The principle of DSSS is that in additional to the traditional modulation, specific spreading codes are used to modulate the baseband data. Figure 4.8 depicts the spreading operation with DSSS.

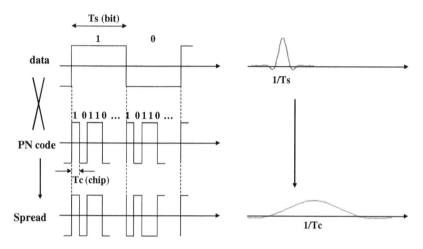

Fig. 4.8 Principle of DS spreading.

The minimum time unit of a DSSS code, at which 0 and 1 alternate, is called chip and is denoted by T_c. Duration of chip T_c is much smaller than a data bit interval. The latter is denoted by T_S in Figure 4.8. Generally, T_S/T_c is called spread factor, which is identical to processing gain for DS. As can be seen in Figure 4.8, the signal spectrum spreads T_S/T_c times after spreading.

Good spreading code must have sharp autocorrelation as defined in Equation (4.22), which shows the similarity of a code with itself of a time delay. Moreover, the cross-correlation defined in Equation (4.23), which describes the similarity between two codes, should be as small as possible.

$$R_{cc}(k) = \sum_{n=0}^{N-1} C_n C_{n+k}^*, \qquad (4.22)$$

$$R_{cg}(k) = \sum_{n=0}^{N-1} C_n G_{n+k}^*, \qquad (4.23)$$

where N denotes code length and * denotes the complex conjugate. Pseudo-noise (PN) sequence is well used as direct sequence spreading code. PN sequence can be generated using maximum length shift register and is referred to as M-sequence [19].

4.3 Ultra-wideband Technologies

UWB has gathered tremendous attentions from both academic researches and industrial developments. UWB benefits a large variety of applications due to the availability of huge frequency spectrum and inherent precision time resolution. On one hand, UWB provides a superior PHY solution for wireless personal area network (WPAN) that requires high data-rate links within short distances. On the other hand, UWB enables high precision ranging with low cost devices, which is essential for wireless sensor networks with positioning requirements. There are two types of UWB technologies in forming UWB signal, i.e., orthogonal frequency division multiplexing (OFDM) and impulse radio (IR). The former is adopted by WiMedia Alliance [9] and the international standard ISO/IEC 26907-2007 for high data-rate link. The latter is adopted in the IEEE standard IEEE 802.15.4a-2007 for low data-rate link as well as precision ranging. Because in most cases, WBAN doesn't require high data rate especially for medical and healthcare applications, IR-type UWB is more favorable. As will be shown in Chapter 6, the required data rate for WBAN is usually between 10 kbps and 10 Mbps in standardization. This is much smaller compared to that defined by WiMedia Alliance.

4.3.1 OFDM-Type UWB

OFDM modulation has been widely adopted in integrated services digital broadcasting for terrestrial (ISDB-T), mobile communications (Mobile WiMAX, LTE), WLAN (IEEE 802.11a), etc., because of its capability of high bandwidth efficiency and robustness against narrow band noise and channel fading. OFDM is a multiple-carrier modulation. A complex OFDM signal can be denoted as:

$$u(t) = \sum_{n=0}^{N-1} C_n d_n e^{j2\pi \left(n - \frac{N}{2}\right) f_0 t}, \tag{4.24}$$

where C_n is a constant of amplitude and d_n is the complex representation of data given by:

$$d_n = a_n + jb_n. \tag{4.25}$$

In Equation (4.24), f_0 is the interval between adjacent carriers. In other words, each carrier frequency is the product of an integer with f_0, i.e.,

$$f_n = n \cdot f_0 (n = 1, 2, \ldots, N - 1). \tag{4.26}$$

Denoting a symbol interval by T, T and f_0 hold the relation of $T = 1/f_0$. It is easy to verify that sine waves satisfy the above conditions are orthogonal to each other. That is,

$$\int_0^N \cos(2\pi m f_0 t) \cos(2\pi n f_0 t) dt = \begin{cases} \dfrac{T}{2}; & (m = n) \\ 0; & (m \neq n) \end{cases}, \tag{4.27}$$

and

$$\int_0^N \sin(2\pi m f_0 t) \sin(2\pi n f_0 t) dt = \begin{cases} \dfrac{T}{2}; & (m = n) \\ 0; & (m \neq n) \end{cases}, \tag{4.28}$$

and

$$\int_0^N \cos(2\pi m f_0 t) \sin(2\pi n f_0 t) dt = 0. \tag{4.29}$$

A conceptual illustration of OFDM signal is shown in Figure 4.9. By sampling Equation (4.24) with a rate of $1/Nf_0$, we obtain:

$$u_k = u\left(\frac{k}{Nf_0}\right) = \sum_{n=0}^{N-1} C_n d_n e^{j2\pi \left(n - \frac{N}{2}\right)\frac{k}{N}}. \tag{4.30}$$

Equation (4.30) is exactly the inverse discrete Fourier transform (IDFT). Introduction of IDFT into OFDM modulation was first proposed in 1971 [10]. It has been becoming popular since 1990s benefiting from the advance of microelectronics. The modulation and demodulation of OFDM are then identical to the operations of IDFT and DFT. A block diagram of an OFDM transceiver is depicted in Figure 4.10. Because the IDFT and DFT are parallel operations, before and after IDFT or DFT, serial-to-parallel and parallel-to-serial conversions must be performed as can be seen in Figure 4.10. The main drawback of OFDM is the large peak-to-average power ratio (PAPR) as shown in Figure 4.9. A number of methods to improve PAPR have been proposed in [11–14].

The OFDM-type UWB defined by WiMedia adopts a tone spacing (f_0) of 4.125 MHz and totally 128 tones (sub-carriers) are required to generate a UWB signal. That results in a UWB signal bandwidth of 528 MHz.

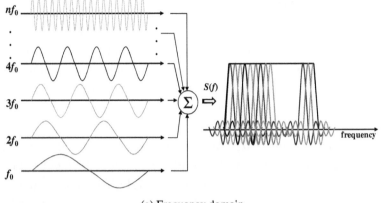

(a) Frequency domain

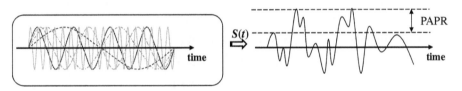

(b) Time domain

Fig. 4.9 A conceptual illustration of OFDM signal.

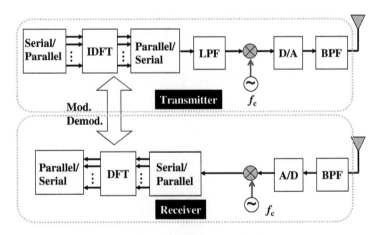

Fig. 4.10 A general diagram of OFDM transceiver.

4.3.2   IR-Type UWB

IR-type UWB coincides with the original concept of UWB, i.e., to employ an ultra-short-time pulse (in nanosecond order) and occupy an ultra-wide bandwidth in frequency domain (more than 500 MHz), for communications [15–18]. We use the conceptual sketches of UWB pulses in Figure 4.11 to explain some basic parameters of IR-type UWB.

Suppose the width of a time pulse is T_p, the occupied bandwidth in frequency domain is about $1/T_p$. An important parameter of IR-type UWB is the pulse repetition frequency (PRF). PRF represents the number of pulses occurring in one second. Pulse repetition interval (PRI), or repetition period, is the inverse of PRF. In an IR-type UWB system, there are usually definitions of peak PRF and average PRF. In Figure 4.11, the peak PRF may be as large as $1/T_p$, while the average PRF is much smaller.

Denoting the average power and peak power of an IR-UWB signal by $P_{average}$ and P_{peak}, respectively, the following relation holds:

$$P_{peak} = P_{average}/\text{PRF}. \tag{4.31}$$

When the UWB signal is restricted by peak power, $P_{average}$ can be increased by using a large PRF. Other advantages of using a large PRF include short acquisition time to detect the signal and capability of providing high data rates. In contrast, the advantages of using a small PRF are low complexity, low cost, and capability of providing good anti-multipath performance.

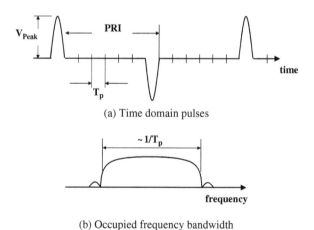

(a) Time domain pulses

(b) Occupied frequency bandwidth

Fig. 4.11 Conceptual sketches of UWB pulses.

If the UWB signal is restricted by the -41.3 dBm/MHz PSD, $P_{average}$ can be calculated as:

$$P_{average} = 10\log_{10}\left[\int_{f_1(\text{MHZ})}^{f_2(\text{MHZ})} s(f)^2 \cdot 10^{-41.3/10} df\right], \qquad (4.32)$$

where $s(f)$ is the UWB signal spectrum, which is dependent on the UWB pulse. Some generally used pulse shapes are Gaussian pulse, root raised cosine pulse, and cosine pulse. In the specifications of IEEE 802.15.4a-2007, a number of pulse shapes, such as Chirp on UWB pulse, continuous spectrum pulse, and linear combination of pulses, are defined to satisfy various requirements. If a UWB signal presents a rectangular spectrum with a bandwidth of $BW(\text{MHz}) = f_2(\text{MHz}) - f_1(\text{MHz})$, Equation (4.32) is simplified to:

$$P_{average} = -41.3 + 10\log_{10}[BW(\text{MHz})]. \qquad (4.33)$$

In general senses, the basic modulations and demodulations used in narrow band can be applied to IR-UWB. To demodulate IR-UWB signals, it is also possible to use coherent, differentially coherent or non-coherent detections.

We show a simple transceiver structure for IR-type UWB in Figure 4.12, in which, an on-off keying (OOK) modulation is employed in transmitter and a non-coherent power detection is adopted in receiver. In the transmitter, a pulse generator is controlled by the controller to send or not to send pulses. The generated pulses are sent with a center frequency of 4.1 GHz. The measured

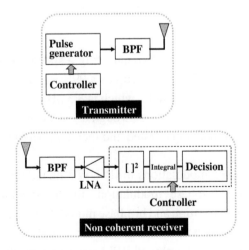

Fig. 4.12 OOK transmitter and non-coherent receiver for IR-UWB.

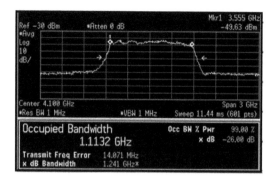

Fig. 4.13 Signal spectrum of generated IR-UWB.

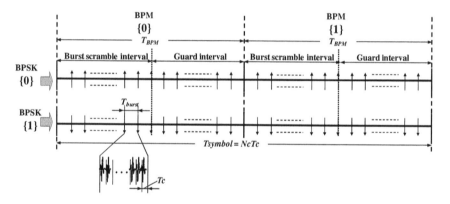

Fig. 4.14 Combined BPM and BPSK modulation.

spectrum of the generated UWB signal using a spectrum analyzer is shown in Figure 4.13. The UWB signal occupies a 99% bandwidth of 1.1 GHz and the PSD is about −49.6 dBm/MHz, which is 8.3 dB lower than the FCC PSD mask. In the receiver, after passing through a low noise amplifier (LNA) the power of received signal is integrated and then compared with a threshold to make decisions.

Other usually used modulations in IR-UWB include pulse position modulation (PPM) and bi-phase modulation. In the specifications of IEEE 802.15.4a-2007, both position and bi-phase modulations are employed to form a combined burst position modulation (BPM) and BPSK modulation. We use Figure 4.14 to explain the combined BPM and BPSK modulations.

Denoting the length of a UWB symbol by T_{symbol}, a total of N_c chips are included in one symbol, i.e.,

$$T_{symbol} = T_c \cdot N_c, \qquad (4.34)$$

where T_c is the length of a chip. One symbol is divided into two equivalent BPM intervals. Each has a length of $T_{BPM} = \frac{1}{2} T_{symbol}$ and carries the data "0" or "1", respectively (position modulation). Furthermore, each BPM interval is divided into two equivalent time intervals. The two intervals are, respectively, referred to as burst scramble interval and guard interval with the same time length of $\frac{1}{4} T_{symbol}$. The former is used for pulse burst hopping to avoid or mitigate collision of pulse bursts from different user devices. The latter is used to absorb the delay spread due to multipath transmission.

The length of a pulse burst is dependent on data rate and is composed of multiple chips. Generally, $T_{burst} \ll T_{symbol}$ so that the burst scramble interval can support multiple users. Moreover, by assigning data '0' and '1' according to the pulse polarities, BPSK modulation (bi-phase modulation) is realized. Say in other words, we can transmit and receive two bits by using the combined BPM and BPSK modulation. However, this is only true for coherent receiver. Non-coherent receiver can only detect the position bit. This characteristic can be used to build FFD and RFD transceivers as discussed at the beginning of this chapter. A FFD detects both BPM and BPSK bits, while a RFD only detect BPM bit.

It should be noted that the delay spread is much smaller for WBAN compared to that considered in the IEEE 802.15.4a-2007 standard. Therefore, the guard interval in Figure 4.14 may be too redundant for WBAN. Moreover, non-coherent structure is more favorable to make a simple and low cost IR-UWB device. However, when high quality and high data rate are the top requirements, coherent structure works better for the purpose.

4.3.3 Band Plans

The main objective of band plan is for efficient usage of the UWB frequency spectrum. It is desirable to design more channels within the available frequency band so as to enable more simultaneously operated WBANs. The band plan must provide enough bandwidth in each channel to contain the specified UWB signals. From the hardware design point of view, centre frequency of each channel should be easier generated from local crystals with small prime factors.

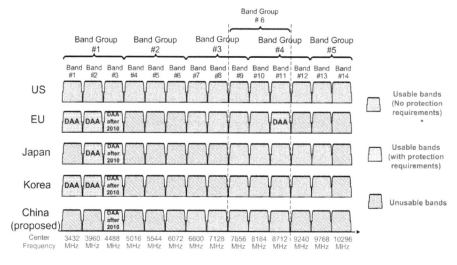

Fig. 4.15 Band plan defined by WiMedia (Copyright 2010, WiMedia Alliance Inc. Used with permission.)

Table 4.1. Band Plan Defined in IEEE 802.15.4a-2007.

Channel number	Center frequency (MHz)	−3 dB Bandwidth (MHz)	Remarks
ch.1	3494.4	499.2	Optional
ch.2	3993.6	499.2	Optional
ch.3	4492.8	499.2	mandatory
w.1	3993.6	1331.2	optional
ch.4	6489.6	499.2	optional
ch.5	6988.8	499.2	optional
w.2	6489.6	1081.6	optional
ch.6	7488.0	499.2	optional
ch.7	7987.2	499.2	mandatory
ch.8	8486.4	499.2	optional
w.3	7987.2	1331.2	optional
ch.9	8985.6	499.2	optional
ch.10	9484.8	499.2	optional
ch.11	9984.0	499.2	optional
w.4	9484.8	1354.97	optional

Two band plans for UWB systems have been defined. One is the band plan defined by WiMedia, which is shown in Figure 4.15 [9]. Another is the band plan defined in IEEE 802.15.4a-2007 which is shown Table 4.1. In WiMedia's band plan, the channel spacing is 528 MHz to contain the spectrum of OFDM UWB, with the intention of to have three channels (#1 − #3) at low band and eight channels (#7 − #14) at high band. In comparison, in

IEEE 802.15.4a's specifications, the channel spacing is 499.2 MHz which is defined at the -3 dB level down from the highest spectrum level. The available channels are three (ch.1–ch.3) at low band and eight (ch.4–ch.11) at high band. In IEEE 802.15.4a's specifications, four channels (w.1–w.4) with large bandwidth are also defined. The large bandwidth channels overlap with the 499.2 MHz channels. The large bandwidth channels are defined for applications with higher quality requirements.

In designing band plan for WBAN, harmonization with IEEE 802.15.4a and/or WiMedia needs to be taken into consideration. When UWB devices from different standards operate in same frequency bands, interference and coexistence become important concerns.

4.4 Channel Coding Technologies

Shannon's channel capacity theorem gives the theoretical limit that a channel can achieve under additive white Gaussian noise (AWGN). Denoting Shannon's channel capacity by C (bits/second), it is given by:

$$C = W \log_2(1 + S/N),$$ (4.35)

where W is the bandwidth of the channel and S/N is the signal-to-noise ratio. Equation (4.35) implies that the channel capacity can be increased steadily with the increase of bandwidth. In practice, spectrum efficiency C/W (bits/second/Hz) is usually used to evaluate a communication scheme, i.e.,

$$\frac{C}{W} = \log_2\left(1 + \frac{E_b}{N_0} \cdot \frac{C}{W}\right).$$ (4.36)

When $C/W \to 0$, we get Shannon's limit of $E_b/N_0 = \ln 2$ $(-1.6$ dB). The more the SNR closes to Shannon's limit to achieve a required BER, the higher the power efficiency is. The purpose of channel coding is to reduce the required SNR, thus to achieve higher bandwidth efficiency and/or higher power efficiency.

4.4.1 Convolutional Codes and Viterbi Decoding

4.4.1.1 Convolutional Codes

Because of the invention of Viterbi decoding [24, 25], convolutional code becomes one of the most popular codes to be used in channel coding in digital communications. Convolutional codes can be used independently or together with other channel coding techniques.

A rate k/n convolutional encoder converts k input binary bits into n output bits by using logical operation (moduo-2 addition) combined with shift registers. A shift register with v memory elements has a constraint length v. When decoding, the computation complexity increases exponentially with v. Although a larger v provides better coding gain, v is usually selected smaller than 9 in practice as a trade-off with decoding complexity.

To get the current n output bits, the logical operation performs not only on the current k input bits but also on forgoing input bits. As an example, a rate of 1/2 encoder with the constraint length 3 is illustrated in Figure 4.16. In Figure 4.16, T_1, T_2, and T_3 denote the three memory elements, respectively. Two output bits d_1 and d_2 are obtained at the two modulo-2 adders when a bit enters the encoder. It can be seen that the current output bits are modulo-2 sum of the current input bit, the bit one time unit ahead (the output of memory element T_2), and/or the bit two time units ahead (the output of memory element T_1):

$$\begin{cases} d_1 = b_1 \oplus b_3 \\ d_2 = b_2 \oplus b_2 \oplus b_3 \end{cases}. \tag{4.37}$$

A convolutional code has different representations. We show in Figure 4.17 a trellis description for the rate of 1/2 trellis code generated by the encoder of Figure 4.16. We denote the four combinations (states) of the two output bits $\{d_1 d_2\}$ by a, b, c, and d and let $a = \{00\}, b = \{01\}, c = \{10\}$, and $d = \{11\}$. In Figure 4.17, a solid line and a dot line, respectively, represent an input bit 0 and 1. Suppose that the initial values at the three memory elements T_1, T_2, and T_3 in Figure 4.16 are 0–0–0, for every input bit 0 or 1, two output bits can be obtained. A solid line (input bit 0) or a dotted line (input bit 1) is used

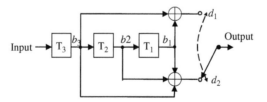

Fig. 4.16 Convolution encoder with constraint length of 3.

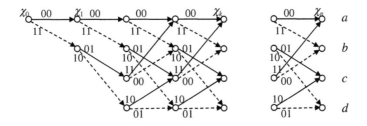

Fig. 4.17 Trellis for the convolutional code of Figure 4.16.

to connect two neighbor nodes. In Figure 4.17, each node represents a state. There are totally four different states a, b, c, and d in the trellis. A solid arrow or a dot arrow is referred to as a branch. On each branch, the values of the two output bits are depicted. Therefore, the trellis diagram in Figure 4.17 describes relationship between input and output of the encoder of Figure 4.16.

4.4.1.2 Viterbi Decoding

We first describe Viterbi algorithm [24] with the trellis of Figure 4.17. A path in the trellis corresponds to a codeword exclusively. A trellis starts from time $i = 0$ and ends at $i = n$ includes all codewords with code length of n. Each single path of length n corresponds to a single codeword and vice versa. There-fore, decoding turns to a problem of finding the shortest path to the received codeword within the trellis. The number of paths in the trellis increases with i. For a code length of n, the total number of paths is 2^n.

Without loss of generality, denoting a node by $\chi_i (i = 0, 1, \ldots, k, \ldots, n,)$, branch metric λ_i is a measure of the branch connecting nodes χ_i and χ_{i+1}. The measure can be Hamming distance or squared Euclidean distance. Then, the path metric between a pair of nodes χ_0 and χ_k can be calculated as:

$$\lambda(\chi_0^k) = \sum_{i=0}^{k-1} \lambda_i. \tag{4.38}$$

At a node χ_k, a path gives the minimum $\lambda(\chi_0^k)$ is called as survival path and it is denoted as $\widehat{\chi}(\chi_k)$. For a trellis with M states (in Figure 4.17, $M = 4$), at an arbitrary node $\widehat{\chi}(\chi_k)$ the number of survival paths is M. In Viterbi algorithm, only the survival paths are used in path metric calculation. Therefore, the calculation complexity is greatly reduced.

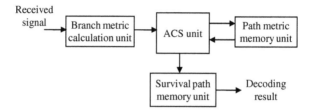

Fig. 4.18 Block diagram of Viterbi decoder.

A block diagram of Viterbi decoder is shown in Figure 4.18. It is composed of a branch metric calculation unit, a path metric add-compare-select (ACS) unit, a path metric memory unit, and a survival path memory unit. The branch metric calculation unit is of the least complexity. At the ACS unit, the current path metric $\lambda(\chi_0^k)$ at node χ_k is decided using the path metric $\lambda(\chi_0^{k-1})$ at node χ_{k-1} and the current branch metric between nodes χ_{k-1} and χ_k. Generally, the ACS unit and the path metric memory unit are combined together to calculate and save path metrics. The survival path unit needs to save M survival paths and for a length of 4 multiply of constraint length or longer.

4.4.2 Block Codes

The beginning of block codes was from the invention of Hamming code [20]. A block code encoder generates n output bits for every k input bits and is denoted as (n, k) code. The n output bits only depend on the current k input bits. Basically, block codes are defined on Galois field (*GF*). A *GF* is a set that is composed of limited elements, on which four basic operations are precedently defined. A *GF* is denoted as $GF(2)$ when binary digits 0 and 1 are the only elements. When the number of elements is q, it is denoted as $GF(q)$. The elements of the extension can be either digits or polynomials. If the number of elements of a *GF* is the exponent of a prime number p, $GF(p^m)$ is called an extension. Denoting the primitive element of $GF(p^m)$ by α, all elements of $GF(p^m)$ can be represented by the exponent of α except for element 0. For examples, $GF(2^2) = GF(4)$ includes four elements of $\{0, 1, x, x + 1\}$, where $x + 1 = x^2$ using the primitive polynomial $p(x) = x^2 + x + 1$.

The most used block codes in digital communications are RS codes and BCH codes as a result of simplicity and performance.

4.4.2.1 BCH code

BCH code is independently invented by Bose and Ray Chaudhuri [22] and Hocquenghem [21]. BCH code is capable of correcting multiple random errors occurred during transmission.

Denoting a codeword of an arbitrary linear block code (n, k) by $\boldsymbol{a} = (a_0, a_1, \ldots, a_{n-1})$, we can cyclically change the position of elements to get a new codeword $\boldsymbol{b} = (a_1, \ldots, a_{n-1}, a_0)$. This linear block code is called cyclic code if and only if the codeword $\boldsymbol{a} = (a_1, \ldots, a_{n-1}, a_0)$ is also a codeword of it. Denoting a block code by polynomial representation:

$$f(x) = a_0 + a_1 x + \cdots + a_{n-1} x^{n-1}. \tag{4.39}$$

The polynomial is called monic if $a_{n-1} = 1$. The least-order monic can be obtained exclusively for cyclic code, which is called generation polynomial. BCH code is a cyclic code and it is defined as follows.

Suppose α is the primitive element of $GF(q^m)$, a cyclic code, which code length is n, the number of elements q, and its generation polynomial $G(x)$ holds roots of $\alpha^l, \alpha^{l+1}, \ldots, \alpha^{l+2t-1}$, is a BCH code that can correct t random errors. The BCH code is called primitive BCH code if $n = q^m - 1$ and l usually takes values of 0 or 1.

Denoting the least-order monic with the root α^i by $M_{\alpha^i}(x)$, the generation polynomial of BCH is given by:

$$G(x) = LCM \lfloor M_{\alpha^l}(x), \ldots, M_{\alpha^{l+1-2t}}(x) \rfloor, \tag{4.40}$$

where $LCM \lfloor \cdot \rfloor$ stands for least common multiple.

As an example, we show a BCH code on $GF(2^4)$ that can correct $t = 2$ random errors. Let $l = 1$, then the generation polynomial hold roots of $\alpha, \alpha^2, \alpha^3$, and α^4. The least-order polynomials on $GF(2^4)$ for α, α^2, and α^4 are the same and given by:

$$M_\alpha(x) = M_{\alpha^2}(x) = \cdots = M_{\alpha^4}(x) = x^4 + x + 1. \tag{4.41}$$

The least-order polynomials for α^3 is:

$$M_{\alpha^3}(x) = x^4 + x^3 + x^2 + x + 1. \tag{4.42}$$

From Equation (4.40), the generation polynomial can be obtained:

$$G(x) = (x^4 + x + 1)(x^4 + x^3 + x^2 + x + 1) = x^8 + x^7 + x^6 + x^4 + 1. \tag{4.43}$$

The data length is the difference between the code length and the highest order of $G(x)$. The code length of this code is $2^4 - 1 = 15$. Thus the data length k is 7.

There are several decoding algorithms proposed for decoding BCH code. Among them, Peterson–Gorenstein–Zierler algorithm, Forney algorithm, and Berlekamp–Massey algorithm are well used. We don't go into details of each decoding algorithm.

4.4.2.2   RS Code

RS code was invented by Reed and Solomon [23]. RS code is a class of non-binary BCH code which maximizes the minimum distance among codewords. One main characteristic of RS code is that it corrects errors in byte. The generation polynomial of RS code on $GF(2^m)$ can be obtained as the least common multiple of the least polynomials of $\alpha^l, \alpha^{l+1}, \ldots, \alpha^{l+n-k-1}$, which are $n - k$ consecutive components of the primitive element α of $GF(q)$, i.e,

$$G(x) = \prod_{i=0}^{n-k-1} (x - \alpha^{l+i}). \tag{4.44}$$

We use RS code on $GF(2^3)$ as an example to illustrate the strategy. The exponent presentation and vector presentation for the elements on $GF(2^3)$ are shown in Table 4.2 for comparison. It can be seen that one element (one byte) of an exponent presentation corresponds to three bits of vector presentation. Let $w_0, w_1, \ldots, w_{n-1}$ represent elements on $GF(2^3)$, the code polynomial can be written as:

$$W(x) = w_{n-1}x^{n-1} + \cdots + w_1 x + w_0. \tag{4.45}$$

Table 4.2.   Exponent and Vector Representations for the Elements on $GF(2^3)$.

Exponent representation	Polynomial representation	Vector representation
0	0	0 0 0
1	1	0 0 1
A	α	0 1 0
α^2	α^2	1 0 0
α^3	$\alpha + 1$	0 1 1
α^4	$\alpha^2 + \alpha$	1 1 0
α^5	$\alpha^2 + \alpha + 1$	1 1 1
α^6	$\alpha^2 + 1$	1 0 1

Suppose that error occurs during transmission and the error polynomial is denoted as:

$$E(x) = e_{n-1}x^{n-1} + \cdots + e_1 x + e_0, \tag{4.46}$$

where e_i ($i = 1, 2, \ldots, n - 1$) are elements of $GF(2^3)$. When $e_i = 0$, no error occurs. When $e_i \neq 0$, error occurs at the corresponding byte W_i. The received signal polynomial can be written as:

$$
\begin{aligned}
Y(x) &= W(x) + E(x) \\
&= (w_{n-1} + e_{n-1})x_{n-1} + \cdots + (w_1 + e_1)x + (w_0 + e_0) \\
&= w_{n-1}x^{n-1} + \cdots + y_1 x + y_0.
\end{aligned}
\tag{4.47}
$$

Denoting the check matrix of an arbitrary linear code as H, the syndrome S is defined as the product of received codeword with the transpose of H.

$$S = YH^T. \tag{4.48}$$

Because for codeword W, it is satisfied that:

$$WH^T = 0. \tag{4.49}$$

We have:

$$S = WH^T + EH^T = EH^T. \tag{4.50}$$

The generation polynomial of RS code on $GF(2^3)$ can be written as:

$$G(x) = (x - 1)(x - \alpha), \tag{4.51}$$

where 1 and α are the roots of the generation polynomial. If error occurs at the i-th byte, syndrome can be obtained as follows:

$$
\begin{cases}
S_1 = Y(1) = e_i \\
S_2 = Y(\alpha) = e_i \alpha^i
\end{cases}.
\tag{4.52}
$$

Solving the equations, we obtain:

$$
\begin{cases}
e_i = S_1 \\
\alpha^i = S_2/S_1
\end{cases},
\tag{4.53}
$$

where e_i and α^i are the error codeword, and the position of the codewords, respectively. By subtracting the error codeword from the received codeword on corresponding position, the error of one byte (three bits) can be corrected.

A simple comparison of BCH codes and RS codes shows that RS codes correct block errors on symbols and usually present better error correcting performance while BCH codes correct multiple bit errors. RS codes have less latency because the processing is in parallel. When complexity is of main concern, BCH codes are favorable because BCH codes require less computation complexity than RS codes do.

4.5 Typical Short-Range Radios

There are some typical technologies enabling short-range radios. ISO/IEC 18092, defined near field communication (NFC), which provides bi-directional communications with a data rate up to several kbps using the 13.56 MHz frequency band. However, NFC only provides a communication range of 10 cm, which is much shorter than the required 3 m in WBAN. Similarly, other examples include specific low level radio and RFID present either limited communication range or limited data rate. In the following, we look at the radio specifications of IEEE 802.15.4 (PHY and MAC for ZigBee) [26], IEEE 802.15.1 (PHY for Bluetooth) [27], and IEEE 802.15.4a (low-rate IR-UWB) for their suitability in supporting WBAN.

The radio specifications of ZigBee (IEEE 802.15.4) are summarized in Table 4.3. Three frequency bands are defined at 868 MHz, 915 MHz, and 2.4 GHz. The first two can, respectively, be used in EU and USA. The 2.4 GHz is available worldwide. The modulation used is OQPSK at 2.4 GHz and BPSK for the other two frequencies. All three frequency bands adopt DSSS. ZigBee provides a data rate up to 250 kbps which is generally available for several 10 m. Therefore, it is able to provide a part of limited solutions for WBAN.

Table 4.3. Radio Specification of ZigBee.

Parameters	Values		
Frequency bands	868 MHz (EU)	915 MHz (USA)	2.4 GHz (Worldwide)
Number of channels	1	10	16
Transmission power	−3 dBm or larger (obeying local regulations)		
Modulations	BPSK	BPSK	OQPSK
Spread spectrum	DSSS	DSSS	DSSS
Chip rate (k chips/second)	300	600	2000
Symbol rate (k symbols/second)	20	40	62.5
Data rate (k bits/second)	20	20	250

Table 4.4. Radio Specification of Bluetooth.

Parameters	Values
Frequency bands	2.4 GHz (Worldwide)
Number of channels	32
Transmission power	0 dBm or larger
Modulations	GFSK
Spread spectrum	FHSS
Data rate (k bits/second)	721 kbps (with extension to 2 Mbps and larger)

However, ZigBee has already been considered mainly for applications for wireless sensor network (WSN). Limited frequency band makes the coexistence between WSN and BAN difficult nevertheless that in ISM band there are also Bluetooth, WLAN, etc.

The radio specifications of Bluetooth (IEEE 802.15.1) are summarized in Table 4.4. The frequency band is 2.4 GHz and modulation is GFSK with FHSS. Communication range is adjustable by changing the transmission power and data rate is 721 kbps with extension to 2 Mbps and larger. Bluetooth provides larger data rate than ZigBee does. In that sense, it is able to support a larger range of WBAN applications than IEEE 802.15.4. However, Bluetooth presents much larger emission power and power consumption than ZigBee does. Those deter its applications to WBAN. It is difficult for Bluetooth devices to operate with a button battery for long time. Moreover, Bluetooth faces the same interference and coexistence problem as ZigBee does. Considering the multiple standards sharing the ISM band including ZigBee, Bluetooth, WLAN, etc., interference among these systems is a complex issue. As consequences, reliability and QoS may be lowered.

IEEE 802.15.4a-2007 can also be a candidate for WBAN. The mandatory radio specifications of UWB in IEEE 802.15.4a are summarized in Table 4.5. The UWB defined in IEEE 802.15.4a is favorable for its small emission power density and small power consumption. The mandatory date rate of 851 kbps and optional data rate can be up to 26 Mbps with a communication range of 30 m or larger. However, IEEE 802.15.4a-2007 is a standard that leans to ranging applications. Both PHY and MAC include numerous features to facilitate ranging such as long preamble, special ranging head, etc. Moreover, the modulation structure may be too redundant for WBAN as discussed in Section 4.3.2. All these make IEEE 802.15.4a less efficient for WBAN.

Table 4.5.   The Mandatory Radio Specifications of UWB Defined in IEEE 802.15.4a.

Parameters	Values
Center frequency	4492.8 MHz
Bandwidth	499.2 MHz (-3dB)
Pulse shape	Square-root raised cosine (roll-off rate $= 0.6$)
Modulations	BPM + BPSK
Channel coding	RS code and convolutional code
Data rate	851 kbps
Average PRF	15.6 MHz or 3.9 MHz

4.6   Summary of Chapter

In this chapter, we overviewed main techniques to implement PHY. The structure of digital communication systems and link budget calculation were first illustrated. Then description emphasis was put on modulations and channel coding. For narrow band modulations, only PSK-type and FSK-type modulations are described because they present small amplitude variation. This is of essential importance for WBAN to achieve good power efficiency. Some of the mentioned modulations are adopted in the PHY of ZigBee and Bluetooth. For UWB modulations, both OFDM type modulation and IR-type modulations are introduced, while the latter is considered to be more suitable for WBAN. For channel coding, traditional trellis codes and block codes are described. They are also good candidates for WBAN in the sense of simplicity and performance.

Implementation dependent parameters, like power consumption, form factor, and cost, are not heavily included. However, all of them are important parameters in practice and must be thoroughly considered.

References

[1]   A. W. Astrin, H.-B. Li, and R. Kohno, "Standardization for body area networks," *IEICE Transactions on Communications*, vol. E92-B, no. 2, pp. 366–372. Feb. 2009.

[2]   Roy L. Ashok, et al., "Next-generation wearable networks," *IEEE Computer Magazine*, pp. 31–39, Nov. 2003.

[3]   F. Graichen, R. Arnold, A. Rohlmann, and G. Bergmann, "Implantable 9-channel telemetry system for in vivo load measurements with orthopedic implants," *IEEE Transactions on Biomedical Engineering*, vol. 54, no. 2, pp. 253–261, Feb. 2007.

[4]   J. G. Proakis, *Digital Communications*, McGraw-Hill International Book Company, 1983.

[5]   P. D. Torino, *Digital Transmission Theory*, Prentice-Hall, Inc., 1987.

[6]   P. Z. Peebles, Jr., *Digital Communication Systems*, Prentice-Hall, Inc., 1987.

[7] G. L. Stuber, "*Principle of Mobile Communication*, Second ed.," Kluwer Academic Publishers, 2001.

[8] D. Divsalar and M. K. Simon, "Multiple-symbol differential detection of MPSK," *IEEE Transactions on Communications*, vol. 38, no. 3, pp. 300–308, March 1990.

[9] http://www.wimedia.org/en/about/ourmembers.asp.

[10] S. B. Weinstein and P. W. Ebert, "Data transmission by frequency-division multiplexing using the discrete Fourier transform," *IEEE Transactions on Communications*, COM-19, pp. 628–634, 1971.

[11] G. Ren, H. Zhang, and Y. Chang, "A complementary clipping transform technique for the reduction of peak-to-average power ratio of OFDM system," *IEEE Transactions Consumer Electronics*, vol. 49, no. 4, pp. 922–926, Nov. 2003.

[12] T. Jiang, W. Xiang, P. C. Richardson, J. Guo, and G. Zhu, "PAPR reduction of OFDM signals using partial transmit sequences with low computational complexity," *IEEE Transactions on Broadcasting*, vol. 53, no. 3, pp. 719–724, 2007.

[13] L. Wang and C. Tellambura, "Analysis of clipping noise and tone-reservation algorithms for peak reduction in OFDM systems," *IEEE Transactions on Vehicular Technology*, vol. 57, no. 3, pp. 1675–1694, 2008.

[14] M.-J. Hao and C.-H. Lai, "Precoding for PAPR reduction of OFDM signals with minimum error probability," *IEEE Transactions on Broadcasting*, vol. 56, no. 1, pp. 120–128, 2010.

[15] D. Porcino and W. Hirt, "Ultra-wideband radio technology: potential and challenges ahead," *IEEE Communications Magazine*, vol. 42, no. 7, pp. 66–74, July 2003.

[16] G. R. Aiello, "Challenges for ultra-wideband (UWB) CMOS integration," in *Microwave Symposium Digest, 2003 IEEE MTT-S International*, vol. I, pp. 8–13, June 2003.

[17] K. Siwiak and D. Mckeown, *Ultra-wideband Radio Technology*, John Wiley & Sons, Ltd., 2004.

[18] M. Ghavami, L. Michael, and R. Kohno, *Ultra Wideband Signals & Systems in Communication Engineering*, John Wiley & Sons, Ltd., 2004.

[19] M. K. Simon, J. K. Omura, R. A. Scholtz, and B. K. Levitt, *Spread Spectrum Communications Handbook*, McGraw-Hill, Inc., 1994.

[20] R. W. Hamming, "Error detecting and error correcting codes," *Bell System Technical Journal*, vol. 29, pp. 147–160, 1950.

[21] A. Hocquenghem, "Codes correcteurs d'erreurs," *Chiffres*, vol. 2, pp. 147–156, 1959.

[22] R. C. Bose and D. K. RayChaudhuri, "On a class of error-correcting binary group codes," *Information and Control*, vol. 3, pp. 68–79, March 1960.

[23] I. S. Reed and G. Solomon, "Polynomial codes over certain finite fields," *Journal of the Society of Industrial and Applied Mathematics*, vol. 8, pp. 300–304, June 1960.

[24] A. J. Viterbi, "Error bounds for convolutional codes and as asymptotically optimum decoding algorithm," *IEEE Transactions on Information Theory*, vol. IT-13, pp. 260–269, April 1967.

[25] D. G. Forney. Jr., "The Viterbi algorithm," *Proceedings IEEE*, vol. 61, pp. 268–278, March 1973.

[26] J. A. Gutierrez, E. H. Callaway. Jr., and R. L. Barrett, Jr., "Low-rate wireless personal area networks — enabling wireless sensors with IEEE 802.15.4™," IEEE Press, 2004.

[27] D. Rosener, "Introduction to BlueTooth engineering," John Wiley & Sons Inc., 2007.

5

Medium Access Control

Medium access control (MAC) is to coordinate multiple users to share a limited amount of radio spectrum resource simultaneously so that the information gets through from a source to a destination. The sharing of the spectrum resource can be in the time domain, frequency domain, code domain, and space domain. It is required to achieve high capacity and fairness by allocating the available bandwidth among multiple users. It should be energy efficient for ad hoc networks. Besides, reliability and quality of service (QoS) are emphasized for medical applications.

5.1 Introduction

Efficient allocation of bandwidth to users is a key design aspect of MAC since the radio spectrum is scarce and expensive. Applications with continuous traffic and delay constrains, such as voice and video, traditionally require dedicate channels to guarantee the expected QoS is not interrupted. Dedicate channels for different users are obtained by orthogonal or non-orthogonal division of signal dimension in the time, frequency, and/or code domain. The MAC schemes divide the total signal dimension into channels and allocate these channels to different users. As shown in Figure 5.1, the most commonly dedicated channel access methods include frequency division multiple access (FDMA), time division multiple access (TDMA), code division multiple access (CDMA), and hybrid of them. Antenna array which makes directional antenna adds an additional space domain. This is called space division multiple access (SDMA). On the other hand, bandwidth allocations

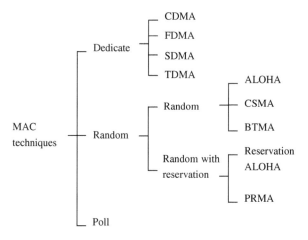

Fig. 5.1 MAC schemes.

for applications with burst traffic are generally random. This is called random access, which typically cannot guarantee the expected QoS always. Because the traffic is generated at random instances, dedicated channel allocation is uneconomical. Moreover, the total number of users (some of them may be in sleep mode) is usually large than the available channels. Figure 5.1 also shows popular random channel access: ALOHA and Carrier sense multiple access (CSMA). In pure ALOHA, the device transmits a packet immediately when the packet is formed. Collision occurs if two or more packets overlap partially or entirely in the same domain. If we neglect the capture effect, any overlap of them destroys the successful reception of all packets. The vulnerable time is two times of the packet duration. By synchronizing packet transmission align with a time boundary; the partial overlap can be avoided. This is called slotted ALOHA. The packet collision can be further reduced if the ongoing transmission is known by the others. In CSMA, each user senses the channel and delays transmission if the channel is detected to be occupied by any others. Then the user conducts random backoff before the next channel sensing, which is to avoid collision of multiple transmissions as soon as the channel is free. To be effective, the channel sensing time and propagation delay must be much smaller than the packet duration. However, the CSMA only works well when all users can effectively detect each other's transmission. Owing to severe attenuation of wireless channels and versatile propagation path, users may be hidden from each other. This is known as hidden node issue, where a busy channel is sensed as an idle one. Instead of the nodes contending the channel spontaneously, the nodes may wait for channel schedule from a master or coordinator device in the star topology network. This is known as poll, which the node get grant to visit channel only when they are asked. The busy tone multiple access (BTMA) replies on centralized star network. When a coordinator device senses the transmission from a device, it broadcasts a busy tone signal to keep all other devices from access the channel. However, the busy tone must be in a dedicate control channel. The random access methods like ALOHA or CSMA work well with short burst traffic. If the traffics are long string of packets or continuous stream data, they work poorly as most of the transmission results in collisions or long time capturing of channel by a single user, which is unfair to others. The scheduled protocols reserve channels for different nodes to avoid conflict with others. Some forms of random access are still needed because a predefined scheduling mechanism is not available at startup. The channel reservation for all subsequent transmission is obtained through contention. Reservation ALOHA and packet reservation multiple access (PRMA) combine the benefit of random access to short burst data with scheduling for continuous data.

In general, the choice of MAC schemes will depend on applications, the traffic characteristics, the required performance (QoS, reliability, and power consumption), the characteristic of underlying channels, and interference from other systems.

Traditional applications like voice and video are continuous in traffic generation. A dedicated channel allocation can facilitate good performance since the transmission is not interrupted by others. This is like circuit switch to establish a unique logic channel for a single user. However, traffic in most data applications is generated at random time instances. A dedicated channel allocation is extremely low efficiency. Random access strategies are used to allocate channel to users that need them. Ethernet is the first popular computer networking technology for local area networks, which was originally over a shared coaxial cable acting as a broadcast transmission medium. The Ethernet adopted CSMA/CD (collision detection) to share the channel in the following steps:

- Data frame ready for transmission.
- Is channel idle? If not, wait until it becomes free.
- Start transmission.

- Is there a collision during transmission? If yes, go to collision detected procedure.
- Reset retransmission counter and end frame transmission.

With the wired medium of Ethernet, e.g., twisted pair and fiber optical, all nodes can "see" each other directly even when they are speaking because separate channels are used for sending and receiving. However, this ability is usually not available any more in the wireless systems, particularly for those systems in industrial, scientific and medicine (ISM) bands and medical bands where the uplink and downlink work in the same frequency channel. When a node is transmitting, it switches off the reception part since usually the local signal is much powerful than signals from the remote node. In other words, collisions cannot be detected while occurring at the sending node. CSMA/CA is used to improve CSMA by not allowing wireless transmission of a node if another node is transmitting.

- Data frame ready for transmission.
- The node listens to the channel for a predetermined amount of time to determine whether or not the channel is being occupied by another node within the wireless range.
- If the channel is sensed idle, then the node starts transmission process.
- If the channel is sensed busy, the node defers its transmission for a random period of time and listens to the channel again.

The action of listening to the channel for a predetermined amount of time, which is usually a short period, is channel sensing or clear channel assessment (CCA). The core idea is to avoid collision before transmission. The CSMA/CA works pretty well for narrowband wireless systems, although the well-known "hidden node" issue exists. However, the transient and carrier-less nature of its low radiation pulses and its harsh multipath channel conditions make channel sensing in the short period extremely unreliable in impulse UWB systems. Almost all nodes become "hidden" to each other. The CSMA retreats to ALOHA.

For wireless sensor networks (WSN), maximizing network lifetime is a major object because the sensor nodes are dead when they are out of battery [1, 2]. There are several reasons for energy waste of a node. The first reason is collision, which means more than one packet arrives at a node during the duration of packets. All collision packets have to be discarded without considering the capture effect. The second reason is overhearing, which means that a node receives packet destined to others or listens to the channel. The third source for energy waste is signaling overhead. A minimum number of signaling packets should be used for a data transmission. Idle listening which is to listen to an idle channel in order to receive a possible packet is the fourth reason. A node may turn on reception circuit early than the transmission of the expected packet. The corresponding reason for transmitter is over-transmitting which means the transmission of a packet starts when the receiver is not ready. The energy efficient MAC protocol should prevent the above energy waste. The IEEE 802.15.4 is a typical example MAC for WSN.

When WSN is used for medical purpose, the reliability and QoS become more important besides energy efficiency. The lost or corruption of medical information due to wireless network may be serious to a patient. Reservation-based TDMA protocols outperform contention-based protocol in providing guaranteed channel access. Recently, Omeni et al. designed a TDMA-based protocol for single-hop wearable wireless body area network (WBAN) in master–slave architecture [3]. Once joining a network, all communications are initiated by the master and slave sensor is uniquely addressed. Therefore, there is no packet collision. Sensor nodes set wakeup fallback time before entering standby mode or sleep mode to reduce power consumption. Otal et al. presented cross-layer fuzzy-rule scheduling algorithm and energy-aware radio activation policies. The key idea is distributed queuing, which divides the TDMA slot into access subslot that is contention based and data subslot that is collision-free [4]. A cross-layer fuzzy logical scheduler takes link quality, packet waiting time, residual battery into consideration to demand or refuse the next collision free data slot. However, its performance is unknown in the fast varying WBAN channels because of the overhead used to maintain the cross-layer information. Marinkovic et al. present similar TDMA frame protocol as Omeni for continuous electroencephalography (EEG) monitor. Particular extra slots are reserved for re-transmission [5].

5.2 Coexistence of WPAN and WLAN for Medical Purpose

The medical environment is a diverse workspace, which encompasses everything from the patient admission process, to examination, diagnosis, therapy, and management of all these procedures. There is desire to use WLAN and WPAN technologies in the unlicensed ISM bands as a common communication infrastructure [6, 7]. The former is typically used for office-oriented applications and patient connection to the outside world, while the latter is usually used for wearable sensors around patient to collect vital information [7–10].

The use of complementary heterogeneous WLANs and WPANs in ISM band in the integrated medical environment brings coexistence, interference, and spectrum utilization issues into picture. Adaptive frequency hopping was proposed for Bluetooth devices to avoid interference from WLAN [11]. A model for analyzing the effect of 802.15.4 on 802.11b performance was provided by Howitt and Gutierrez [12]. The degradation of WLAN performance is small given that the WPAN activity is low. However, the high duty cycle of WLAN traffic can drastically affect the WPAN performance [10]. A distributed adaptation strategy for WPAN based on *Q-learning* has been proposed to minimize the impact of 802.11 interference [13]. However, the spatial reuse issue in the heterogeneous networks has not drawn much attention. Some researches have shown that the spatial reuse and aggregate throughput in the homogeneous WLAN mesh network are closely related to physical channel sensing. Xiao and Vaidya showed that the aggregate throughput can suffer significant loss with an inappropriate choice of carrier sense threshold [14]. Zhu et al. reported that a tunable sensing threshold can effectively leverage the spatial reuse [15]. Zhai and Fang found that the optimal carrier sensing threshold for one-hop flows does not work for multihop flows [16]. Ramachandran and Roy showed cross-layer dependence between carrier sense and system performance [17]. However, the impact of carrier sense has not been studied in the heterogeneous networks. Simulators widely used for performance evaluation, like NS-2 and OPNET, do not contain detailed physical layer module like carrier sense. For the lack of carrier sensing knowledge between WPAN and WLAN, Golmie et al. simply simulated two carrier sensing cases: the WPAN can only detect packets of its own type and the WPAN can also detect WLAN's transmission, in their coexistence study for medical applications [10].

The IEEE 802.15.4 and 802.11b can be considered as examples in the heterogeneous integrated medical environments [2, 3, 5]. Table 5.1 lists the system parameters of them. In the table, thermal noise is -174 dBm per hertz, 8 dB implementation losses, and 8 dB radio noise figure were assumed. Only four WPAN channels locate in the guard band of WLAN. Both of them adopt

Table 5.1. WLAN and WPAN parameters.

Parameters	802.15.4	802.11b
Transmission power (dBm)	0	16
Channel bandwidth (MHz)	2	22
Background noise (dBm)	−94.9	−84.6
Spread code (chips)	32/4 bits	11
Data rate (Mbps)	0.25	11
CCA window (μs)	128	15

CSMA-CA protocol. CCA which detects an incoming packet and ensures a free channel before transmission is an essential element of the CSMA protocol. The CCA processes received radio signals in a suitable time termed CCA window. It then reports channel state, either busy or idle, by comparing the detection with a threshold. The CCA can be either energy based, or feature based, or a combination of two. The energy-based CCA integrates signal strength from radio front end during the CCA window; while the feature-based CCA looks for the known features, e.g., the modulation and spreading characteristics, of the signal over the channel. Modulated signals are in general coupled with sine wave carriers, pulse trains, repeating spreading, or cyclic prefixes, which result in built-in periodicity. This periodicity can be used to detect signal of a particular modulation type. The feature-based CCA performs far better than the energy-based CCA. However, a prior knowledge of the signal characteristic is needed. And the CCA module needs a dedicated detector for every potential coexistence signal class. The main advantages of energy-based CCA are its simplicity, generality, and low power consumption. It is a universal mechanism and can be deployed in various systems. This is particularly useful for the Cognitive Radio (CR) in the white space. Unlike feature-based CCA, there is no need for waiting time for the specific features of the signal [17]. The downside of energy based CCA is that it is prone to false detection.

We applied only energy-based CCA to 802.11b and 802.15.4 in the coexistence environments because there have been nearly ten IEEE wireless technologies with different modulations, band plans, and transmission powers operating at 2.4 GHz ISM bands. The low cost medical sensors based on 802.15.4 are unpractical to have all the knowledge for feature-based CCA, which is usually complex and power-hungry [17].

5.2.1 Energy-Based CCA

In mathematics, CCA is a test of two hypotheses:

$$\begin{cases} H_0 : y[n] = w[n] & \text{signal absent} \\ H_1 : y[n] = x[n] + w[n] & \text{signal present} \end{cases} \tag{5.1}$$

where $x[n]$ is the targeted signal, $w[n]$ is the white Gaussian noise with variance σ^2, and $n = 1, \ldots, N$ is the sample index in total N-independent samples. Under common detection performance criteria, e.g., Neyman–Pearson criteria, likelihood ratio yields the optimal hypothesis testing solution. CCA metric is compared to a threshold Γ to make a decision. The CCA performance is characterized by a pair possibility: detection possibility, P_d, and false alarm possibility, P_{fa}.

For simplicity, assuming that the energy-based CCA is realized by a non-coherent module that integrates the square of the received signals in baseband and sums its samples in either analog or digital domain. In particular, the energy detector can be a quadrature receiver with y_I and y_Q representing samples of signal on the I (in-phase) and Q (quadrature) branches respectively. The energy-based CCA metric can be given by:

$$Y = \sum_{n=1}^{N} (|y_I[n]|^2 + |y_Q[n]|^2). \tag{5.2}$$

In the AWGN channel, each $|y_I[n]|$ and $|y_Q[n]|$ has a normal distribution with mean μ and variance σ^2 and Y can be evaluated as generalized Chi-square function $Y \sim \chi^2(\lambda, 2N)$, where $2N$ is the degree of freedom and $\lambda = \sigma^2 + \mu^2$. Under the H_0 hypothesis, each normal distribution has $\mu = 0$. Thus, Y has a χ^2 distribution. Under the H_1 hypothesis, in the presence of signals with an SNR $\gamma = \frac{\mu^2}{\sigma^2}$, Y has a non-central χ^2 distribution. The mean and variance are as follows [13]:

$$\begin{cases} H_0 : \mu_0 = \sigma^2, & \sigma_0^2 = \dfrac{2}{N}\sigma^4 \\ \\ H_1 : \mu_1 = \mu^2 + \sigma^2, & \sigma_1^2 = \dfrac{2}{N}(2\mu^2 + \sigma^4). \end{cases} \tag{5.3}$$

When N is large, using central limit theory, the energy-based CCA metric in Equation (5.1) can be approximated as Gaussian random process. Then P_d

and P_{fa} can be expressed in terms of the Q function:

$$P_d = Q\left(\frac{\frac{\Gamma}{\sigma^2} - (1 + \gamma)}{\sqrt{\frac{2}{N}(1 + 2\gamma)}}\right), \quad P_{fa} = Q\left(\frac{\frac{\Gamma}{\sigma^2} - 1}{\sqrt{\frac{2}{N}}}\right). \quad (5.4)$$

Theoretically the energy-based CCA can meet any desired P_d and P_{fa} simultaneously if the number of samples in CCA window is unlimited. Given a limited N, the CCA ability is obviously determined by the *SNR* of the signal. There is an inherent tradeoff between P_d and P_{fa}. Define the CCA error floor as sum of the two errors at an optimal threshold, which can be found by equating $1 - P_d$ and P_{fa}. Using Equation (5.4), the CCA error floor is obtained:

$$P_{CCA\_ef} = Q\left(\sqrt{N}\frac{\gamma}{1 + \sqrt{1 + 2\gamma}}\right). \quad (5.5)$$

Note that the error floor depends on the number of symbol chips and *SNR*. When *SNR* $\ll 1$, Equation (5.5) can be approximated as:

$$P_{CCA\_ef} = Q\left(\sqrt{N}\frac{\gamma}{2}\right). \quad (5.6)$$

A linear decrease in *SNR* requires a quadratic increase in N to maintain the same error floor.

5.2.2 Asymmetric CCA Error Floor

Table 5.2 lists the numbers of signal chips in the CCA windows of IEEE WLAN and WPAN. Figure 5.2 shows the CCA error floor in the heterogeneous networks environment as per Equation (5.5) The systems are assumed to be only equipped with modules to receive the signal of its own type. The error floors decrease with increment in signal chips in the CCA window. Given the defined CCA windows, the CCA abilities, e.g., sensitivity and operation range, to determine channel state are different. This is termed asymmetric

Table 5.2. Number of signal chips, N, in the CCA window.

Sensed signal device	802.15.4 signals	802.11b signals
802.15.4	32*8	128*11
802.11b	15*2	15*11

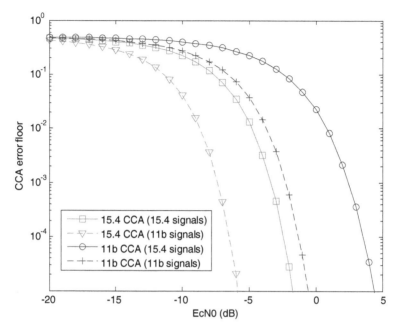

Fig. 5.2 Error floor of energy-based CCA in integrated medical environment (AWGN channel).

CCA. Under the same SNR conditions, the lowest error floor is the WPAN to sense 802.11b signals; the highest error floor the WLAN to sense 802.15.4 signals. The performance difference is nearly 10 dB.

The asymmetry CCA error floor can be attributed to differences in the underlying signals over channel (power, symbol rate, and background noise) and CCA window. In physics, a higher data rate and a longer CCA window means more signal pulses in baseband can be collected. Better CCA performance is a natural result. Asymmetric CCA error floor can be further reinforced by other factors in the integrated medical environment:

- difference in transmission powers which is usually stronger for WLAN.
- difference in channel bandwidth which are 22 MHz and 2 MHz for the WLAN and WPAN.

For both WPAN and WLAN, the performances to detect the signal of its own type are similar. There is no big difference in the numbers of signal chip in the CCA window.

Table 5.3. SNRs (dB) to achieve 1‰ communication BER and CCA error floors in AWGN channel.

Device Signal	802.15.4	802.11b
Communication	−0.8 dB	5.6 dB
802.15.4 CCA	−3.2 dB	2.6 dB
802.11b CCA	−7.2 dB	−2.2 dB

Table 5.3 compares communication ability with CCA ability when both are set to an error probability of 1‰. As expected, the CCA range is larger than the communication range. For WPAN, sensing 802.11b signals has 4 dB greater link margin compared to sensing the signals of its own type. In contrast, for WLAN sensing 802.15.4 signals require a 4.8 dB higher SNR.

5.2.3 Energy-Based CCA in Integrated Medical Environments

Asymmetric CCA error floor indicates channel sensing insensitive or oversensitive to other signals in the coexistence of WLAN and WPAN environment. The asymmetric CCA in the heterogeneous networks is different from the traditional "hidden node" or "exposed node" issues in the homogeneous network. In the homogeneous network, two devices belonging to the same system are reciprocal in their ability to sense each other. However, in the heterogeneous networks, the sensing abilities of different systems are unequal and depend on the underlying signals over channel and the separation distances. As shown in Figure 5.2, WLAN signals are well sensed by both of them, but WPAN signals could be ignored by the WLAN systems when they are separated by enough space.

Looking at Equation (5.4), in both the H_0 and the H_1 hypotheses, the energy distribution is related to N. This means that the CCA threshold is related to the underlying physical signals. Figure 5.3 depicts the distribution of collected energy over a free channel based on different underlying signal assumptions. When WLAN senses channel per data rate of WLAN, the collected channel energy is small due to a short integration time. However, more collections can be obtained during the CCA window. When WLAN senses channel per data rate of WPAN, stronger channel energy with fewer samples is obtained due to a long integration time. The energy distributions shown in Figure 5.3 indicate that the optimal CCA thresholds quite depend on the type of underlying signals. Prior knowledge is, therefore, needed to optimize the CCA performance.

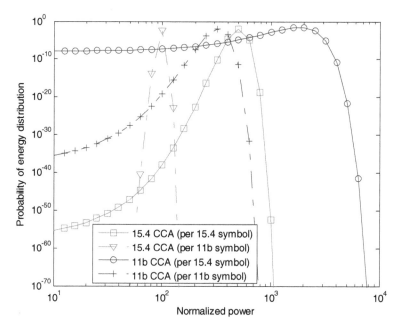

Fig. 5.3 Collected power distributions of energy-based CCA over a free channel per different underlying signals.

Usually, the Neyman–Pearson criteria are adopted in CCA because a miss-detection of a busy channel is riskier than a false alarm of a free channel. Equation (5.4) can be re-written as:

$$P_d = Q\left(\frac{\sqrt{4N}\,Q^{-1}(P_{fa}) - 2N\gamma}{\sqrt{4N(1 + 2\gamma)}}\right). \tag{5.7}$$

As expected, P_f is independent of γ since there is no signal under H_0 hypotheses. However, the CCA threshold Γ is optimized for its own type of signals, not for other signals, because the devices have no ideas on the heterogeneous networks environment in prior. The "non-optimized" Γ may result in CCA that is insensitive or oversensitive to other types of signal. In the extreme case, an opposite channel state could be obtained. This is unilateral sensing, where device A can sense the activities of device B, but not *vice versa*. Assuming that the selected SNR corresponds the doubled distances to achieve BER of 0.1‰ for WPAN, and the SNR were measured by the signal type of its own [19]. The receiver operating characteristics (ROCs) of energy-based CCA were plotted in Figure 5.4, in which we set SNR by −9.5 dB. As mentioned above, for

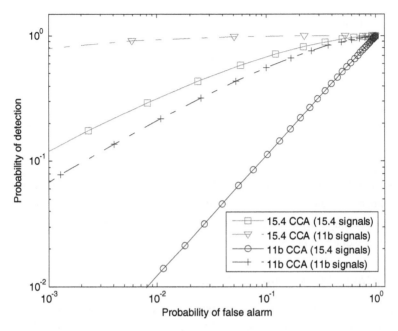

Fig. 5.4 ROC of energy-based CCA in integrated medical environment (SNR = −9.5 dB measured by WPAN in AWGN channel).

WPAN sensing, the 802.11b signals offer the best detection. The WLAN's sensing of the 802.15.4 signals, in contrast, is prone to fail at such a low SNR. A particular reason for the worse performance is the mismatch of channel bandwidths. When WLAN applies a 22 MHz bandpass filter to WPAN signals, an additional 10.4 dB noise is introduced.

Until now we consider the AWNG channel. When the channel is varying due to fading and shadowing, Equation (5.7) gives the CCA performance conditioned on the instantaneous γ. The average CCA performance can be derived by averaging Equation (5.7) over fading statistics:

$$P_d = \int_0^\infty Q\left(\frac{\sqrt{4N}\,Q^{-1}(P_{fa}) - 2N\gamma}{\sqrt{4N(1 + 2\gamma)}}\right) f(\gamma)d\gamma, \qquad (5.8)$$

where $f(\gamma)$ is the probability of distribution function (PDF) of SNR under fading. The medium-scale variance of SNR can be characterized by log-normal distribution [19]. The log-normal shadowing is usually described in terms of its dB-spread, σ_{dB}, which is related to σ by $\sigma = \sigma_{dB} \ln(10)/10$. Under

Rayleigh fading, the SNR γ is Rayleigh distributed:

$$f(\gamma) = \frac{1}{\bar{\gamma}} \exp\left(\frac{\gamma}{\bar{\gamma}}\right) \quad \gamma \geq 0, \tag{5.9}$$

where $\bar{\gamma}$ denotes to average SNR. Because it is difficult to have a close-form expression of Equation (5.8) over fading channels, it is evaluated numerically.

Figure 5.5 plots the ROC of energy-based CCA over log-normal shadowing and Rayleigh fading channel. Comparing with the AWGN curves, a degraded CCA performance can be observed and the degradations are closely related with the CCA parameters and SNR. In other words, meeting the desired performance demands a longer CCA window in the fading channel. However, in practice it may not have reliable detection even with an infinite CCA window in the presence of noise uncertainty [20].

5.2.4   Impact of Asymmetric CCA

Table 5.4 lists the required minimum SNRs and their corresponding distances to achieve reliable energy-based CCA ($P_{FA} < 1\%$ and $P_D > 90\%$). The corresponding distances were computed using Equations (5.1)–(5.3) and the parameters listed in Table 5.1. For WPAN, the sensing of 802.11b signals is reliable at an SNR as low as -9.25 dB. This SNR is 9.65 dB lower than the critical SNR for communication. The critical SNR for communication is the least SNR to achieve BER $< 0.1\%$. The CCA range is 180 m longer than the communication range. In contrast, sensing 802.15.4 signal by WLAN requires a high SNR up to 9.75 dB, which is 3.15 dB more than the critical SNR for communication. The CCA range is 42 m shorter than the communication range.

Figure 5.6 qualitatively compares the communication range and CCA ranges in the integrated medical environment. We can define a "heterogeneous exclusive CCA range" (HECR), in which systems in the heterogeneous environment can reliably sense the activities of each other. In the considered scenario, the HECR is the maximum distance that WLAN can sense 802.15 signals. Given the system parameters and assumptions, the HECR for IEEE WLAN and WPAN is 25 m. Good coexistence between them can be expected when they are located within the HECR. However, it becomes different when they are separated beyond the HECR. For WPAN systems, the CCA range of WLAN signals is more than twice as long as the communication range. And it is longer than the CCA range of its own signal type. That is, the WPAN is

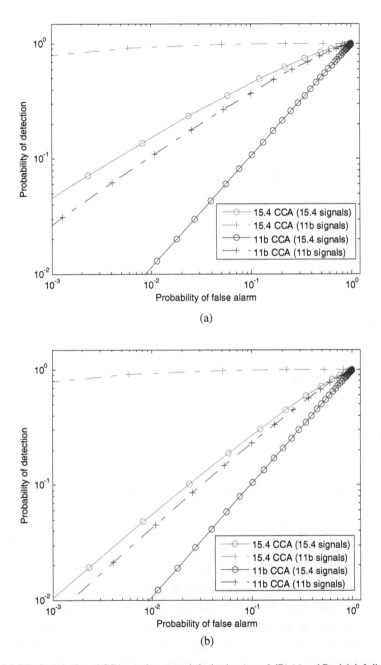

Fig. 5.5 ROC of energy-based CCA over log-normal shadowing ($\sigma = 6$ dB) (a) and Rayleigh fading (b).

Table 5.4. Minimum SNRs (dB) and corresponding distances (m) to achieve CCA (PFA < 1%, PD > 90%).

	802.11b CCA		802.15.4 CCA	
	WPAN signals	WLAN signals	WPAN signals	WLAN signals
SNR (dB)	9.75	−4.5	−6	−9.25
Corresponding distance (m)	25	200	155	280

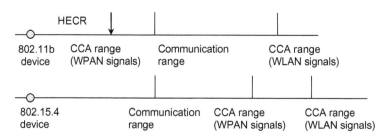

Fig. 5.6 Communication range and CCA ranges in integrated medical environment.

oversensitive to the WLAN signals. It can even sense a WLAN packet that is outside of the keep-out range of receiver in the worst case. The keep-out range defines the minimum separation at which WPAN and WLAN do not interfere each other. Although the oversensitive CCA avoids the "hidden node" issue, it suffers from the "exposed node" issue. This results in poor spatial reuse of frequency channels and low aggregation throughput since WPAN sometimes unnecessarily withdraw packet before transmission. As simulated in Ref. [16], the threshold optimized to maximize aggregate throughput is higher than the optimal threshold for a single hop. For WLAN systems, the CCA range of WPAN signals is about a quarter of the communication range. Packet collision may occur when WLAN traffic occurs later than the WPAN traffic.

The concept of HECR distinguishes from most coexistence studies which usually assumed that WLAN cannot sense the activities of WPAN [10, 12, 13]. Although the HECR of 25 m is not sufficient for outdoor applications, it is good enough for most indoor bedside medical applications defined by IEEE 1073 are within this range. However, putting oversensitive and insensitive CCAs together results in an unfair share of channel between WLAN and WPAN when they are separated beyond HECR. There is a preferential treatment of WLAN traffic. The WLAN is overprotected, while the WPAN is vulnerable. Therefore, it is necessary for both systems to distinguish whether the medium is occupied by signal of its own type or the other signals and have different CCA thresholds associated with the underlying signals. For WPAN systems, a higher CCA threshold to sense 802.11b signals can increase the spatial re-usage; while for WLAN systems, a lower CCA threshold to sense 802.15.4 signals can improve the fairness of medium sharing.

5.3 Dynamic On-body WBAN Channels

Understanding the on-body channel is the first step toward achieving a successful design of WBAN systems, especially for the MAC design. One of the most significant features of the on-body channel is that the propagation path between two points on the human body is not stationary [21]. Even when individuals stand still, their bodies are subject to small movements, such as breathing. These movements become significant during normal activities and sports. As a result, the on-body channel becomes time-varying under dynamic scenarios in which the whole body or part of the body is in movement, such as running and walking [22]. Significant channel fading has been found in associated with the movement. The fading pattern may impact the MAC design, for example, the arrangement of re-transmission once error occurs.

Researchers have studied the dynamic on-body channel in terms of interference and multipath that combine with the transmitted signal to produce a distorted waveform at receiver [21–24]. However, the temporal stochastic characteristic of the dynamic on-body channel has not been well studied. The level crossing rate (LCR) and average fading duration (AFD) can only provide the first-moment statistics of the time-varying features [23]. The combined effect of time-varying fading, path loss, and noise of the on-body channel results in a discrete channel with memory, in which the error occurs in clusters separated by error-free gaps. Finite-state Markov chain (FSMC) is the preferred mathematical tool to describe the burst behavior of time-varying channels [25]. A generalized Gilbert–Elliott model partitions channel per its quality into a finite number of intervals, each of which corresponds to one state of the channel [26–30]. The simplest Gilbert channel has two states that correspond to the absence of errors and error occurrence with a defined probability, respectively [26–28]. Example criteria include bit or packet error rate, packet error distribution, packet throughput, and fade duration [25]. The single-error-state Fritchman model has been successfully used to describe the high-frequency channel and the fading effect in mobile radio channels. Another benefit is that the FSMC-based methods allow system performance evaluation over flat fading channels in closed form.

5.3.1 Relative Path Loss in Dynamic WBAN Channel

As shown in Figure 5.7, measurements in Refs. [31, 32, 39, 40] considered 10 Rx positions in three scenarios in 4.5 GHz. The standing still scenario served as reference scenario, and two dynamic scenarios: walking in place and standing up and down on the chair, involve human movements. The distances between the Tx antennas and the Rx antenna were given in Table 5.5. At each position, the peak value of the channel impulse responses was considered as the path loss of the measured instant. For each Rx antenna position in the reference scenario, all path losses measured in the session were averaged to obtain a reference path loss. The instant path loss at the position in the two dynamic scenarios was normalized by the reference path loss to obtain relative path loss in dB. The relative path loss can partially remove the impact of distance between the Tx antenna and the Rx antenna at different positions in the reference scenario.

Figure 5.8 plots the channel response in the frequency domain obtained at the right wrist when the subject was walking. The plot clearly shows that the fading that results from movement is non-frequency selective in the studied

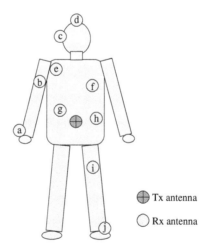

Fig. 5.7 Locations and movement on the human body (A: right wrist, B: right upper arm, C: right ear, D: head, E: shoulder, F: chest, G: right rib, H: left waist, I: thigh, J: ankle).

Table 5.5. Statistical parameters for the dynamic on-body channels ($\theta = -10$ dB).

Positions	Distance (mm)	Walking Relative path loss (dB)	Standing up/down Relative path loss (dB)
A	440–525	-3.99 ± 11.8	8.45 ± 3.48
B	360	4.45 ± 2.50	6.30 ± 5.67
C	710	-4.15 ± 2.47	-3.13 ± 5.25
D	650	-3.54 ± 3.21	-3.69 ± 3.21
E	310	-1.80 ± 2.05	0.22 ± 4.10
F	230	3.64 ± 2.47	5.08 ± 6.41
G	183	-0.89 ± 1.40	5.70 ± 3.21
H	140	-1.50 ± 0.98	-3.56 ± 3.35
I	340	-2.69 ± 2.65	-0.60 ± 3.77
J	815–940	0.94 ± 3.70	-1.13 ± 4.72

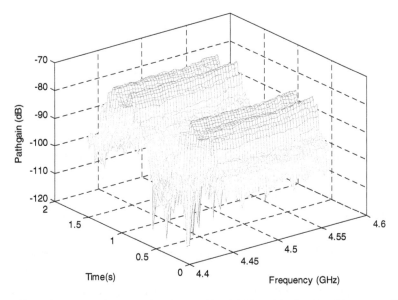

Fig. 5.8 Channel response in frequency domain at position A (right wrist) when the subject was walking.

frequency band. Figure 5.9 shows the cumulative distribution function (CDF) of the relative path loss which is the path loss in the two dynamic scenarios normalized by the reference path loss at position A for each action. The body movement produces shadowing-like effects as the movement cause changes in the separation, orientation and operation of antennas. In some cases, the

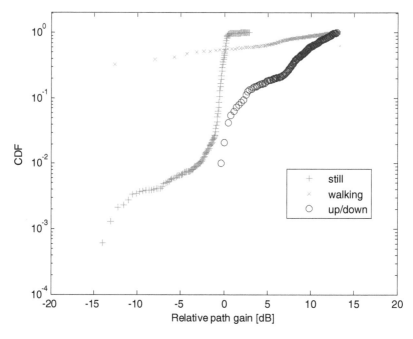

Fig. 5.9 CDF of the relative path loss at position A (right wrist).

movements may introduce obstacles between antennas; this changes the relative path gain significantly. A larger range of relative path gain can be found in the two dynamic scenarios. Table 5.5 lists the statistical parameters, mean value and standard deviation, of the relative path gain at each Rx antenna position.

Figure 5.10 draws time variation of the relative path loss at position A in different scenarios. The channel experiences significant fading that follows the subject motion in a regular base. In the reference scenario as well, involuntary movements of the subject cause irregular and abrupt fading at some instants. The time-varying on-body channel is extremely sensitive to body movements. The statistic parameters of the relative path gains at each Rx antenna position are also in Table 5.5.

By combining all measurements at 10 antenna positions together, a general concept of on-body propagation which is independent of antenna position and dynamic scenario can be obtained. Figure 5.11 shows the probability distribution function (PDF) of relative path losses in each action.

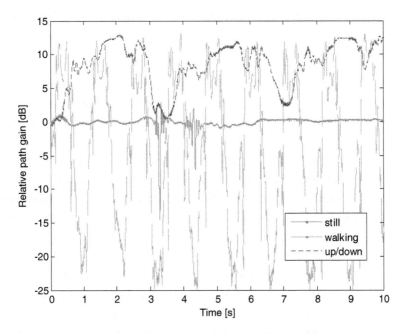

Fig. 5.10 Time-varying relative path loss at position A (right wrist).

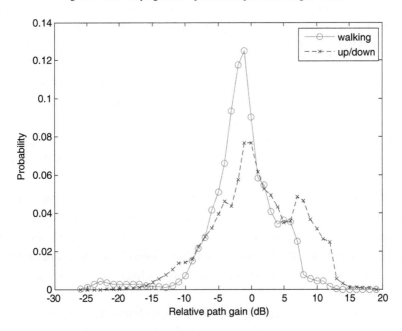

Fig. 5.11 PDF of relative path loss.

5.3.2   Statistical Analysis of the On-Body Dynamic Fading Pattern

Medical sensors usually prefer simple modulations without complex error coding for the reason of complexity, cost, and power consumption. The performance of simple modulations deteriorates significantly when the SNR is lower than a threshold value, θ, which corresponding to a fading margin. Such errors are commonly referred to as bursts. When signal quality is above the threshold, it is usually an error-free communication. We term the channel as good when the relative path gain is above the threshold θ, and *vice versa*.

The burst characteristic can be described by the LCR, which is the frequency of a signal that crosses a threshold in a positive or negative going direction, and the AFD, which is the total time for a signal dwells below the same threshold. Figure 5.12(a) and (b) plots the average LCR and AFD for the dynamic scenarios. Table 5.6 lists the AFD for each action and position for $\theta = -10$ dB. The threshold was selected since it was typically the least link margin for a system design. The AFD reveals that the on-body fading characteristic is position dependent and action dependent. The position G is always in good channel during both actions, while positions A, B, E, F, and H experience bad channel during only one of actions. The other positions have bad channel experience in both actions. Totally the channel is bad in 4.83% of time.

The AFD only provides the first-moment statistics of the time-varying features. It is interesting to study the distribution of channel dwelling time above or below a threshold to give more information of the randomness. We drew CDF of durations of bad channels and good channel for three different thresholds in Figure 5.13(a). Table 5.6 gives more details about the duration of channels at each position. In all cases, it is observed that the dominant bad channels are short duration. For example, at $\theta = -10$ dB the duration of approximately 85% of bad channels are shorter than 10 ms. This confirms that the on-body channel is highly sensitive to body movements. A small movement, such as the introduction of obstacles between antennas, may have a significant impact on the path loss. For lower values of θ, the duration of the bad channels are observed to be smaller. None of the bad channels last for more than 400 ms. The maximum duration of channels is related to velocity of movement.

As shown in Figure 5.13(b) most good channels are shorter than 10 ms. This is similar to the bad channels. However, the good channels in short duration

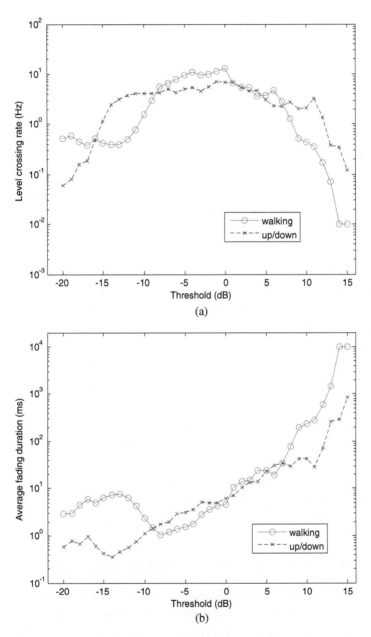

Fig. 5.12 AFD in two dynamic scenarios.

Table 5.6. AFD and LCR of the dynamic on-body fading ($\theta = -10$ dB).

| Positions | Walking | | | Standing up/down | | |
| | Level crossing | Duration (ms) | | Level crossing | Duration (ms) | |
		Good channels	Bad channels		Good channels	Bad channels
A	31	204.23 ± 252.37	99.03 ± 155.97	0	10,000	0
B	0	10,000	0	16	423.33 ± 851.80	12.6 ± 21.98
C	60	159.03 ± 509.90	1.65 ± 2.68	204	37.56 ± 171.64	7.15 ± 25.70
D	37	67.64 ± 251.40	2.83 ± 3.75	123	62.06 ± 265.87	10.23 ± 43.57
E	0	10,000	0	3	1666.7 ± 250.6	1.0 ± 5.8
F	0	10,000	0	17	572.56 ± 779.69	14.12 ± 18.13
G	0	10,000	0	0	10,000	0
H	0	10,000	0	8	862.71 ± 1174.41	51.88 ± 71.44
I	9	750.35 ± 1403.24	2.00 ± 2.45	14	710.08 ± 840.69	25.15 ± 26.94
J	18	574.12 ± 459.97	10.82 ± 4.75	30	262.48 ± 538.37	20.17 ± 47.19

(<20 ms) are not sensitive to the threshold θ. There is almost no difference for the three thresholds. More than 10% good channels last longer than 400 ms. In other words, the channel may dwell in good channel for a long time. Figure 5.14 shows position-dependent channel dwelling time using positions C and J as examples.

Given $\theta = -10$ dB, it is interesting to see how bad channels and good channels interleave each other in Figure 5.15. The horizontal axis is the duration of the bad channels, and the vertical axis is the duration of the good channel immediately after it. Large amount of short good channels (<20 ms) are after bad channels. These abrupt changes probably result from involuntary movements, like breath or antenna rotation. Even small movements may lead to mismatch of antenna's direction and deep fading. The standing up/down leads to more abrupt channel changes than walking. This is because that it involves movements of more parts of body comparing with walking.

5.3.3   5-State Fritchman Model for On-Body Fading Pattern

The Fritchman model was first introduced in 1967. The simplest Gilbert channel has two states that correspond to the absence of errors and error occurrence with a defined probability, respectively [26]. For binary channels, Fritchman's framework divides the state space into k error-free states and $N - k$ error states according to the signal-to-noise ratio (SNR) at receiver [27–29]. The measured data shown in Table 5.6 and Figure 5.13 show that the on-body channel dwells

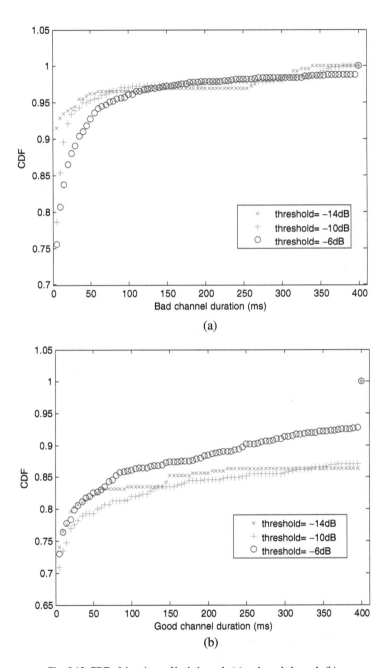

Fig. 5.13 CDF of durations of bad channels (a) and good channels (b).

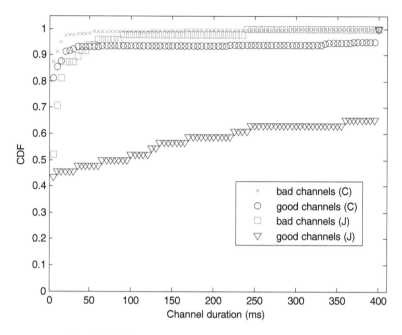

Fig. 5.14 CDF of channel dwelling times at positions C and J.

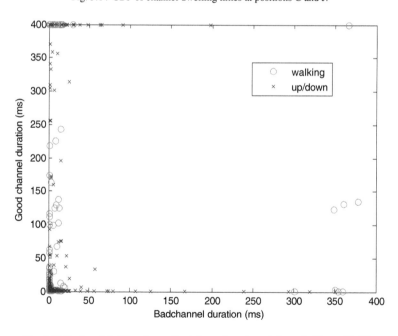

Fig. 5.15 Scatter plot of durations of bad channels and good channels ($\theta = -10$ dB).

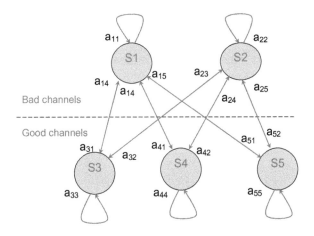

Fig. 5.16 A 5-state Fritchman model for dynamic on-body fading pattern.

in good channel and bad channel with different patterns. From the viewpoint of coding and medium control, it is more interesting to know the interleave pattern of channels in different SNRs shown in Figure 5.15 so as to correct the communication errors. Therefore, the channel dwelling pattern should be considered for determining the state as well as the SNR. Figure 5.16 shows a five-state Fritchman model for describing the burst behaviors of on-body channels:

- S1: unstable error-free state, good channels which last less than 20 ms.
- S2: semi-constant error-free state, good channel which are over 20 ms and less than 400 ms.
- S3: constant error-free state, good channel which are over 400 ms.
- S4: unstable error state, bad channel which last less than 20 ms.
- S5: semi-constant error state, bad channel which are less than 400 ms.

The state of on-body channel at time t is given by:

$$\Pi_t = \Pi_{t-1} A \quad t \geq 1, \tag{5.10}$$

where $A = \begin{bmatrix} a_{11} & 0 & 0 & a_{14} & a_{15} \\ 0 & a_{22} & 0 & a_{24} & a_{25} \\ 0 & 0 & a_{33} & a_{34} & a_{35} \\ a_{41} & a_{42} & a_{43} & a_{44} & 0 \\ a_{51} & a_{52} & a_{53} & 0 & a_{55} \end{bmatrix}$ is the state transit matrix and $\prod_0$ is the

initial state probability vector. The error generation matrix takes a very simple form:

$$B = \begin{bmatrix} 1 & 1 & 1 & 0 & 0 \\ 0 & 0 & 0 & 1 & 1 \end{bmatrix}. \tag{5.11}$$

Table 5.7 lists statistical parameters of channel in different states for $\theta = -10$ dB. By categorizing good channels and bad channels into different states per their durations, we applied Baum-Welch algorithm to estimate the parameters in Equation (5.11) [33]. Table 5.8 shows results for different actions. Comparing two actions, the biggest difference is in S1. The walking action dwells in S1 more time, and the action of standing up/down is more easily to transit to S4. By selecting corresponding data, both action-independent and position-dependent state transit matrix can be obtained.

A 200-s segment of dynamic on-body fading was generated using values listed in Table 5.7. The channel state was calculated every 1 ms. Figure 5.17 compares the CDF of the simulated channel dwelling times ($\theta = -10$ dB) with that of the measured ones. The initial state was S1. A good fit between them was obtained. Comparing Figure 5.17(a) with (b), the walking action leads to more bad channels with short duration, but fewer good channels with short duration. This is in agreement with the results in Table 5.6.

Table 5.7.   Statistical parameters for dynamic on-body fading in different states ($\theta = -10$ dB).

Parameters		State samples		State durations (ms)	
		Walking	Standing up/down	Walking	Standing up/down
Good	S1	93	337	3.31 ± 3.66	2.38 ± 3.05
channels	S2	29	34	99.41 ± 65.15	161.1 ± 132.3
	S3	32	38	2515.5 ± 3342.9	1771.7 ± 2082.4
Bad	S4	145	371	3.45 ± 4.38	3.48 ± 4.20
channels	S5	10	38	287.1 ± 139.3	80.68 ± 87.26

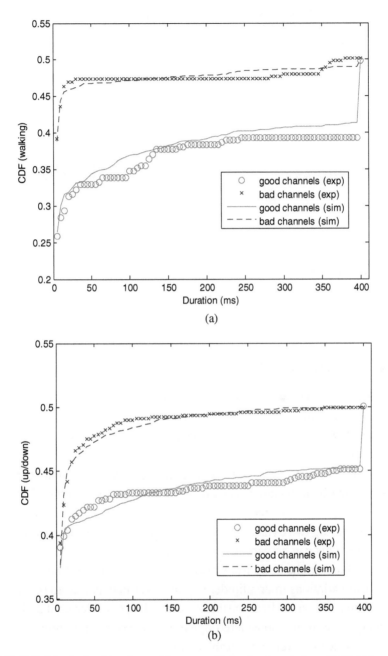

Fig. 5.17 CDF of the simulated durations and the measured fading durations for $\theta = -10\,\text{dB}$, (a) in walking and (b) standing up/down.

Table 5.8. Parameters of the 5-state Fritchman model for different action scenarios.

Threshold $\theta = -10$ (dB)		A and Π_0				

Walking

$$A = \begin{bmatrix} 0.772 & 0 & 0 & 0.223 & 0.005 \\ 0 & 0.991 & 0 & 0.0082 & 0.00087 \\ 0 & 0 & 0.9997 & 0.00025 & 0.00007 \\ 0.137 & 0.047 & 0.043 & 0.773 & 0 \\ 0.00189 & 0.0009 & 0.00063 & 0 & 0.996 \end{bmatrix}$$

$$\Pi_0 = [0.00415 \ 0.0346 \ 0.923 \ 0.0064 \ 0.032]$$

Standing up/down

$$A = \begin{bmatrix} 0.69 & 0 & 0 & 0.29 & 0.019 \\ 0 & 0.993 & 0 & 0.005 & 0.0017 \\ 0 & 0 & 0.9995 & 0.00032 & 0.00016 \\ 0.195 & 0.0209 & 0.0183 & 0.766 & 0 \\ 0.0076 & 0.00147 & 0.0044 & 0 & 0.986 \end{bmatrix}$$

$$\Pi_0 = [0.011 \ 0.058 \ 0.882 \ 0.0158 \ 0.034]$$

Actions combined

$$A = \begin{bmatrix} 0.7131 & 0 & 0 & 0.2715 & 0.01542 \\ 0 & 0.9924 & 0 & 0.006169 & 0.00140 \\ 0 & 0 & 0.9996 & 0.000283 & 0.00012 \\ 0.1783 & 0.0284 & 0.0252 & 0.768 & 0 \\ 0.00486 & 0.00122 & 0.00258 & 0 & 0.9913 \end{bmatrix}$$

$$\Pi_0 = [0.0075 \ 0.046 \ 0.902 \ 0.011 \ 0.033]$$

5.3.4 On-Body Fading Map

It is interesting to map the fading characteristics described by the Fritchman states onto the human body, as shown in Figure 5.18. This again confirms that the on-body fading depends on antenna positions and action scenarios. Positions G and E are totally free from fading. Positions A, B, F, and H experience fading during one of the two dynamic scenarios. Other positions experience fading during both dynamic scenarios. The right panel of Figure 5.18 classified positions per error-free states. Positions A, C, D, and J experience state S4 and S5 during two dynamic scenarios, while other positions always remain in state S3. This can be partially attributed to the fact that all of positions B, F, H, and I are on the body trunk. They therefore underwent relatively small movement variations in amplitude. It is however unexpected to see positions C and D that are on the head sustain severe fading considering that they were relatively static in the considered dynamic scenarios. This might be due to involuntary head movements during actions.

The question is why the short-term good channels and bad channel (less than 20 ms) described by S1 and S4 in the Fritchman models dominate the

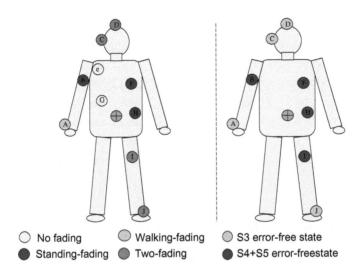

| ○ No fading | ○ Walking-fading | ○ S3 error-free state |
| ● Standing-fading | ● Two-fading | ● S4+S5 error-freestate |

Fig. 5.18 Map of the on-body fading pattern.

dynamic on-body channels? The reason could be that antennas on the body surface are expected to move along with actions. There are mainly two reasons to contribute signal attenuation in the on-body channels. Obviously the first reason is the distance between Tx device (antenna) and Rx antenna. The distance between them changes during action and the changes are position dependent. Most wearable devices are attached or bound on the body surface. When the device moves along with the body during actions, both Tx antenna and Rx antenna may rotate themselves around a point more or less. This is the second reason: mismatch of antenna direction. As well known, an antenna placed on the body surface is heavily influenced by its surroundings [21]. As described in Chapter 3, the omni-directional antenna becomes a directional antenna in the near space of the body.

Relative movements between two on-body antennas can be classified into intentional movements and unintentional movements depending on whether the movement can be controlled or not. The intentional movements are controllable. Consider the distance between Rx position A and Tx position in Figure 5.7 during walking. When a hand is in front of the body, it is shorter than that when it is behind the body. Meanwhile, antenna at Rx position A rotates around the shoulder joint. In contrast, the unintentional movements cannot be controlled. One reason for unintentional movement is the shrinking and expanding of muscles which underlies wearable devices. Due to imperfect mount of wearable devices, rolling mules during action may slightly spin an antenna along an axis. People may also have some unintentional movements even in the standing still, e.g., breath and heart beat. Analyzing the fading pattern and on-body movement, it can be inferred that the middle and long-term patterns described by S2, S3, and S5 can be attributed to the intended movements; the dominant short-term pattern described by S1 and S4 can be attributed to the mismatch of antenna direction resulting from unintentional movements.

5.3.5 Retransmission Strategy for Dynamic On-Body Fading

The communication errors that occur in the bad channel states must be corrected. Because the good states and bad states alternate in the dynamic on-body channel, retransmission after some period of backoff until the channel switching to good state is a natural choice. Usually, the medical sensor can enter an inactive mode to save power by shutting down a part of the circuit. After the backoff period, the medical sensor enters active mode to transmit the packet again. The backoff may further take the pulsed charge conditions of battery to improve the lifetime of the battery [35]. Another reason is that the simple medical sensor cannot afford resources for complex automatic repeat request (ARQ) strategies such as Go-Back-N and Selective Repeat [36, 37].

As shown in Figure 5.19, the probability of a successful retransmission after backoff depends on the current channel state and the packet duration. We considered transmission packets of durations 1 ms and 10 ms. The 10-ms-long packet is longer than most short fading durations induced by body movements. The initial states are S4 and S5 in which communication errors occur. As speculated, the channel in S4 switches to good state more easily after a short backoff than does the channel in S5. A long packet is vulnerable to the dynamic fading channel because the channel must dwell in good states for a longer period of time. Hence, a longer backoff period is needed. It is always preferable to adopt a long backoff period in all the cases because the channel has more probability to switch to and dwell in good states. This coincides with that the channel dwells in good states in most of the time.

For the resource-limited medical sensors, the backoff must be optimized to satisfy the stringent requirements for latency and power consumption. For

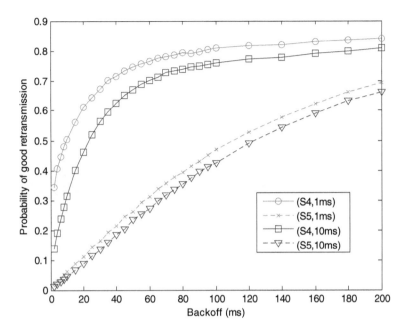

Fig. 5.19 Probability of the first good retransmission.

example, life-critical information must be transmitted with minimum latency. For other real-time or best-effort traffic, power consumption must be considered when the latency requirements are satisfied. Both power consumption and latency are determined by the probability of successful retransmission. Figure 5.20(a) shows the plots of average latency of a good retransmission. In the case of short packets, it is preferable to retransmit the packet immediately because the 1-ms packet duration is within the scope of most short good channels. On the other hand in the case of a long packet, optimal backoff is essential to achieve minimum latency. This is because unnecessary immediate retransmission may segment a good channel that is slightly longer than the packet into two parts. As a result, both the retransmissions may fail. For the same packet duration, there is a big difference in latency whether the error occurs in S4 or S5.

The total power consumption for a good retransmission is the sum of the power consumed in the inactive mode and that consumed in packet transmission. Without loss of generality, we assume that the ratio of power consumption in the inactive and active modes is 1:1000. For typical sensors, the drain current

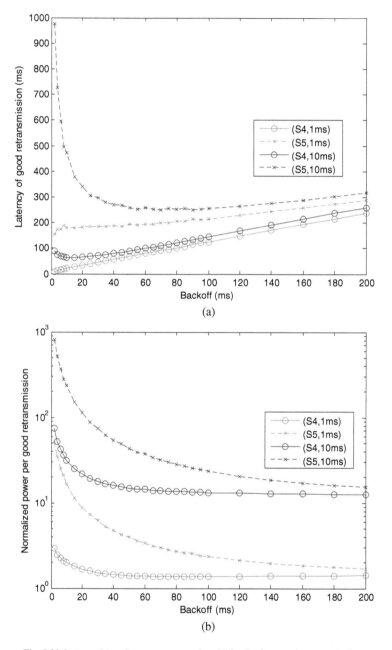

Fig. 5.20 Latency (a) and power consumption (b) for the first good retransmission.

in active mode is in the mA level; while in the inactive mode the drain current is in the μA level. The plots of power consumption for a good retransmission are shown in Figure 5.20(b). The y-axis is normalized to the power consumption for transmission of a single packet. Although there is an optimal backoff period to minimize the power consumption, the advantage of adopting long backoff is less than 10% even when the backoff is doubled. Thus choosing a long backoff period is not a big burden to battery. Moreover, the impact of initial channel state is apparent from Figure 5.20(b).

As seen in Figures 5.19 and 5.20, the optimal backoff required for minimizing the latency and power consumption largely depends on the packet duration and the current channel state in which errors occur. However, we cannot determine the channel state by observing a single packet. Therefore, a channel state probing scheme is necessary.

5.4 Requirements for WBAN for Medical Applications

The WBAN aims to support low complexity, low cost, ultralow power, and highly reliable wireless communications for use in close proximity to, or inside, the human body (but not limited to humans) to satisfy an evolutionary set of entertainment and healthcare products and services.

5.4.1 WBAN for Medical Applications

The medical and healthcare applications distinguish WBAN from general WSN which are mainly implemented by low-rate WPAN technologies [43]. As shown in Table 5.9, they share common features like limited resources, low/modest duty cycle, energy efficiency, plug-and-play, diverse coexistence environments, and heterogeneous device ability. The significant differences lie in the sensor device, dependability, networking, traffic pattern, and channel.

First, WBAN medical applications consider safety, quality, and reliability as top priority, while general WSN are cost sensitive for market reasons. To improve reliability, general WSN usually distribute redundant sensors as backup for sensing, transmission, and forwarding. In contrast, there is little redundancy in medical WBAN for medical reasons. Some vital signals, like electrocardiogram (ECG) and EEG, are location dependent and there is limited area on the body surface. In other words, they can only be measured by

Table 5.9. Comparison between Medical WBAN and General WSN.

	Medical WBAN	General WSN
Common features	Limited resources: battery, computation, memory, energy efficiency	
	Diverse coexistence environment	
	low/modest data rate, low duty cycle	
	Dynamic network scale, plug-and-play, heterogeneous devices	
	ability, dense distribution	
Sensor/Actuator	Single-function device	Multi-function device
	Fast relative movement in small range	Rare or slow movement in
	device lifetime, days, < 10 years	large-range network lifetime and
	(implant sensor)	device lifetime, months,
		< 10 years
	Safe (low SAR) and quality first	Cost sensitive
Dependability	Reliability (first), guaranteed QoS	Expected QoS, redundancy-based
		reliability
	Strongly security (except emergency)	Required security
Networking	Small-scale star network	Large-scale hierarchical network
	No redundancy in device	redundant distribution
	Deterministic node distribution	Random node distribution
Traffic	Periodical real time (dominant), burst	Burst (dominant), periodical
	(priority)	
	Unidirectional traffic	Unidirectional or bidirectional traffic
	M:1 communication	M:1 or point–point communication
Channel	Specific medical channel, ISM band	ISM band
	Body surface or through body	Obstacle is unknown

deterministic location. It makes no sense to allocate sensors outside of the interest/effect area. Thus it is difficult to allocate redundant sensors in the limited area.

Second, the medical WBAN has more frequency bands to select than general WSN which usually work in the ISM bands. Coexistence issue is one of the big concerns in the ISM bands. Although the specific medical bands are less noisy, they are narrowband and conditional license. For example, the WMTS band can only be used in the licensed hospital and clinic, but not at home. FCC allocated a frequency band in 402–405 MHz for MICS on a shared, secondary basis in 1999. As stated in Chapter 2, the MICS rule is quite different from that of ISM band and any other licensed bands.

Third, the traffic pattern of medical WBAN is featured by periodical real time data (e.g., EEG and ECG) and some top priority burst data (e.g., alarm and alert) [43]. In contrast, general WSN typically consider versatile traffic. The medical message, especially the alarm, has very strict requirement in terms of QoS, which are more stringent than the general WSN.

Fourth, the wireless implant communications outstand WBAN from other WSN applications. Medical implants may have more stringent limitation in size and weight, and therefore limited processing, memory, and power capacities. However, lifetime of implant devices which are usually in continuous operation must be maximized to avoid the risk, cost, and patient trauma inherent in replacement surgical procedure. Power management of low power transceiver, processor, and sensor/actuator, and sometimes energy harvest are necessary. Furthermore, the material used should be biocompatibility with human body since human immune system will combat foreign substances in the body. Location of implant is another challenge. A medical implant will be located by physician to where it provides the best patient care and comfort, with little consideration on the radio propagation and network. Figure 5.21 depicts WBAN implants and the implementation concerns [44].

More or less, it also means the existing IEEE 802.15.4b and IEEE 802.15.4a cannot fully support the medical application efficiently and economically [47, 48]. The lack of redundancy, priority traffic, dominant periodical data, guaranteed QoS in versatile coexistence environment and wireless implant communications challenge physical (PHY) and medium access control (MAC) designs of WBAN.

5.4.2 General Description of WBAN

The WBAN addresses medical/healthcare applications and other non-medical applications with diverse requirements. The medical applications include

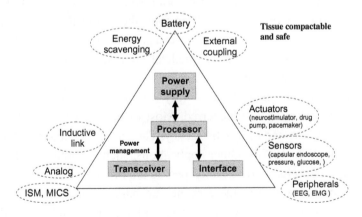

Fig. 5.21 WBAN implant and concerns of implementation.

continuous waveform sampling of biomedical signals, monitoring of vital signal information, and low rate remote control of medical devices, etc. The non-medical applications include video and audio, bulk and small data transfer, command and control for interactive gaming, etc. Depending on the application, WBAN devices may require a network of anywhere from a few sensors or actuator devices communicating to a piconet controller, which can be implemented in a handset, a PDA, or a laptop. In another example, potentially hundreds of sensors and actuators (such as EEG) can communicate to a gateway device through which they are connected to the Internet.

Devices for the above applications are usually highly constrained in terms of resource such as CPU processing power, battery capacity and memory size, and operate in unstable environments, like in the scenario of dynamic action described in the previous section. At the same time, medical sensors or actuators may have to be physically small to be wearable or implantable. However, the gateway device is typically less constrained than the sensors or actuators.

The devices would operate indoor or outdoor in the home, hospital, small clinic, fitness center, etc. There may be interference from and to the other devices in the environment. Patients may simultaneously have both medical applications and non-medical applications, and both wearable and implantable applications on/in its body.

Because of the space limitation and location-dependent characteristics of medical information, it is unlikely to deploy redundant medical sensors for vital information collection. Depending on the philosophy of medical applications, the major traffic tends to be point-to-multipoint (e.g., stimuli) and multipoint-to-point (e.g., ECG). Therefore, the traffic flow is asymmetric. During diagnosis, doctor may investigate a parameter in a command/response mode. The "downstream" traffic (medical commands) is coming from the gateway to a particular sensor or actuator.

Applications are likely to have very different requirements in terms of the amount of data to be transferred. Most of biomedical and vital signals tend to be periodic and of low frequency. The packet generation interval can vary from 1 ms to 1000 s. Other applications, such as motion detection and fall detection for elderly or disabled people are event-triggered and therefore communications related to these events are bursty. Some applications may involve transmitting a log file once a day, with typically kilobytes of data.

Some medical sensors may detect alarm conditions and require prioritization of messages. The raw data may be used to generate alarm conditions. In medical applications this might be detection of heart beat stoppage, excessively low or high blood pressure or body temperature, excessively low or high blood glucose level in a diabetic patient. Another example is the battery dying in a monitor WBAN sensor. Time and reliability crucial alarm packets are expected to have higher priority than sensing data.

Remote medical monitoring and control applications can be "open loop" or "closed loop". In the former case, the sensing data makes its way through a gateway device to the caregiver, who may decide to take an action, and send control information to the actuators in the network. The "closed-loop" control in the future trend will flow over local loops without intervention from the caregiver. Closed-loop control may have a latency requirement that can be 100 ms to seconds. In many of these applications, if the packets do not arrive within the specified interval, the system may enter an emergency alarm state.

5.4.3 Overall Requirements

Given the broad range of possible applications space, key features for the MAC must be to address the issue of scalability in terms of data rates, power consumption, network size, and security [46]. In some cases the tradeoff for speed or security may increase power consumption or indeed for improved quality of service. Other tradeoffs need to be addressed in either or both the MAC or PHY layers. High-level characteristics of the MAC and PHY layers are summarized as follows.

- The network components may be in close proximity to, or inside the human body. Note the special case of the head, which has most of sensory nerves for video, audio, and others.
- The WBAN should be able to recover from link and node failures.
- Typical link data rate should be some tens of kb/s in most of the cases. However, raw data rate can be up to 10 Mbps for some multimedia applications and low data rate can be as low as 10 kbps in some medical applications.
- The power consumption should allow for self-powered operating time without intervention from several hours to several years.

- The QoS support should be provided for most medical and non-medical applications.
- Security shall be energy efficient with minimal overhead and support at least authentication, data integrity, and encryption operations when needed.
- Coexistence between WBAN piconets, coexistence between WBAN and other wireless technologies, and coexistence of WBAN in medical environments, electromagnetic compatibility (EMC) and electromagnetic interference (EMI) should be addressed.
- SAR into the body must satisfy the relevant regulatory requirements.

The MAC protocol should be independent underlying PHYs which include narrowband PHY layer, UWB PHY layer, and MICS PHY layer. The MAC mainly considers the requirements for QoS, power consumption, and coexistence.

5.4.4   Quality of Service Requirements

QoS is an important part in the framework of risk management for WBAN medical applications. The critical factor is the reliability of the transmission, meaning that appropriate error detection and correction methods, interference avoidance methods, or any other suitable techniques should be provided at MAC layer. Other QoS measurements include point-to-point delay and delay variation. QoS provisions should be flexible such that they can be tailored to suit application needs. The QoS parameters have a strong impact on MAC: reliability, latency, and jitter (variation of one-way transmission delay) should be supported for real-time applications that need them. The latency in medical applications should be less than 125 ms. The latency in non-medical applications should be less than 250 ms and jitter should be less than 50 ms. Capability of providing fast (< 1 s) and reliable reaction in emergency situations and alarm message, which have higher priority than others, should be provided after the network has been set up. One such requirement is to transmit an "emergency" condition that the WBAN node has detected. While this "emergency" transmission is in progress, power consumption saving and coexistence can be relaxed. Power saving mechanism (such as duty cycling) should be provided, whilst not impacting application latency requirement.

Channel migration mechanism should be considered to provide the required reliability. Periodical traffic and burst traffic should be effectively supported.

5.4.5   Power Consumption Requirements

Most medical WBAN devices should operate while supporting a battery life of months or years without intervention; whereas other non-medical devices may require a battery life of tens of hours depending on the nature of the applications and/or physical constraints on the size of the devices. For example, cardiac defibrillators and pacemakers are expected to have a lifetime of more than 5 years, whereas swallowable camera pills typically have lifetime of 12 hours. Most of the non-medical applications have stand-by power requirement of 100–200 hours and active power requirement of several hours.

Ultralow power operation is crucial for longevity of implanted WBAN devices. In some applications energy scavenging techniques may be employed which may alleviate the need for a battery.

Peak power consumption and average power consumption should be minimized to support small form factor battery and maximize battery life. It is common for low duty cycle devices to shut down radio and CPU resource for most of the time. Efficient and flexible duty cycling techniques are required to minimize idle listening, overhearing, packet collision and control overhead. In addition the coordination of nodes should not induce frequent wake up of nodes.

5.4.6   Coexistence Requirements

WBAN devices will co-exist with other WBAN devices and legacy devices in the interference environments. Both medical application in hospital, small clinic, healthcare center, and home should be considered, along with wearable entertainment applications. The devices should have attributes to deal with interference ingress (interference coming into the PHY layer) and interference egress (interference caused by the PHY layer). The attributes may be adjusted by higher layer management.

The devices must be able to operate in high noise, high multipath, and dynamic environments. MAC and PHY should work together to support densely co-located operation of multiple WBAN piconets. In particular, implantable WBAN piconet and wearable WBAN piconet should gracefully coexist in-and-around the body. A fair bandwidth sharing among collocated WBAN and graceful degradation of service is desirable for high duty cycle applications, while uncoordinated operation is acceptable for low duty cycle applications. Medical applications may be given higher priority than entertainment non-medical applications when bandwidth is scarce.

5.5   Unified MAC Protocol for WBAN

As described in Chapter 2, different frequency bands are expected to be used by WBAN, a unified MAC mean an MAC protocol which is independent of underlying PHYs. It is particularly useful for industry which aims the global market with different radio regulations and different purposes. Of course, the MAC is a tradeoff on considering multiple features of WBAN some of them may contradict each other, like the QoS and power consumption. Complexity in implementation must also be taken into consideration.

5.5.1   Overview of MAC Protocol

As shown in Figure 5.22, a WBAN piconet consists of a coordinator C, which is the coordinator of the whole network and a number of nodes. The basic architecture of WBAN piconet is the star topology. An optional two-hop extension is supported on considering that the body can sometimes block the signal path. The relay capable node N4 in Figure 5.22, can forward message from the blocked node to coordinator. Both node and coordinator can be on the body or in the near space of body. However, only node can be inside the body. As

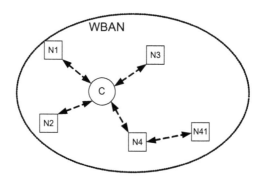

Fig. 5.22 WBAN network topology: star and two-hop extension.

stated in Chapter 2, an in-body coordinator is unpractical in the state-of-art techniques.

A WBAN piconet can operate in either beacon mode where is the beacon message is broadcasted periodically or non-beacon mode where beacon broadcast is not allowed [49]. In the beacon mode, the time is naturally divided into superframe in time axis. The superframe, which is usually bounded by beacon packet, consists of up to 256 time slots with equal duration because the total devices around on and implanted in the body are expected to be less than 256 in the near future for most medical applications [46]. The time slots in a superframe must be integral times of 4 for easy implementation. When the beacon is not allowed or wanted for regulation reason and/or specific application reason, WBAN piconet can operate in the non-beacon mode. There is no superframe concept in the non-beacon mode. MICS systems can only work in the non-beacon mode because the beacon broadcast is not allowed per FCC's MICS rule.

In some WBAN piconets, time reference is preferred to so that a circuit-switch service can be provided. This makes it easy to guarantee the QoS requirement by reserving a fixed allocation or a dynamic allocation to an application. It is also good for power management. The node can wake up at a predefined time. In the beacon mode, beacon is a natural choice to indicate the time reference. However, applications like MICS service which can only operate in the non-beacon mode also require scheduled channel access. A timed-poll (T-Poll) message is used to deliver the time information in the non-beacon mode. Of course, MAC layer can also depend on application layer to provide time reference. In order to keep accurate time information, the node must wake up to listen to the beacon or T-Poll to refresh its clock and time information. This is a major source of power drain. Battery power consumption can be reduced through hibernation and sleep state to shut down most part of its circuit. The hibernation state denotes to a long inactive period which is usually multiple superframes, while the sleep state denotes to the short inactive period within a superframe. Some low duty cycle applications, which have a few transmissions in a day for example, may prefer no time reference to save more battery power. There is no need to wake up and listen to the time reference information.

The WBAN coordinator may adopt different schemes to solve the coexistence and interference mitigation issue within a WBAN piconet between a medical application and a non-medical application and among WBAN piconets. At the piconet level, a WBAN piconet is classified into different categories from life critical medical to entertainment. Channel hoping allows a WBAN to hop to a new channel after dwelling on the current channel for a given time. Beacon is the most important packet to manage the WBAN piconet. Time hopping can help to avoid consecutive beacon collision among different WBAN piconets in the same channel. When all channels are occupied, different WBAN piconets can share the channel cooperatively or non-cooperatively through passive channel.

5.5.2 Beacon Mode

As shown in Figure 5.23, the superframe structure defined in the beacon mode is bounded by regular beacons transmitted by the coordinator [49]. The superframe structure is divided into an active portion and an inactive portion. In the inactive portion, the coordinator and devices may enter low power (sleep) mode for power conservation. The coordinator is responsible for maintaining the timing of superframe structure.

In the active portion when data transmission occurs, the structure is further sequentially divided into the beacon, the priority access period (PAP), the contention access period (CAP), and the contention-free period (CFP). The CAP supports contention-based channel access for one-time prompt traffic, such as association, bandwidth request, and some low duty cycle traffics. The CFP supports scheduled channel access for periodical traffics and QoS guaranteed traffics, such as medical waveforms and stream of audio and video. The PAP supports high priority communications for classified emergency or alarm traffic which is life critical over any other traffic. The superframe is

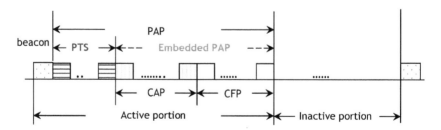

Fig. 5.23 Structure of WBAN superframe.

divided into equal length time slots for easy computing of channel capability given a data rate. This reduces the required processing power for a device.

In the beacon mode, beacon is in charge of managing a WBAN piconet such as coordination of channel access, power control, and maintaining clock synchronization. Although the regular beacon is mandatory, listening to the beacon is node dependant. The node can skip some of them as long as its clock is accurate for the application. In the inactive superframe, the beacon transmission is skipped to save the power consumption of coordinator.

5.5.2.1 Random channel period

The CAP is used for uplink traffic only. In the CAP, devices that wish to transmit will compete for channel access using slotted ALOHA mechanism. The ALOHA, instead of CSMA mechanism which depends on CCA operation, is considered because of the underlying PHY features. As shown in Figure 5.24:

- The impulse UWB signals are hard to be detected due to its weak, transient and carrier-less features, and harsh channel condition.
- Fast attenuation of in-body MICS signals (< 300 mm propagation distance).
- Deep fading of on-body narrowband signals in the case of non-line of sight (NLOS).
- High noise floor of HBC systems.
- Fast and deep fading of on-body system due to human movements.

The above channel conditions make it unreliable for CCA operation. An unreliable CCA results in the well-known "hidden node" issue in the CSMA. Consequently, the CSMA deteriorates to pure ALOHA channel access, which is of worse performance than the slotted ALOHA.

Network reliabilities and packet latencies are critical for medical applications, especially for emergency life-threatening data. In order to ensure most instantaneous emergency data transmissions, WBAN traffics are classified into different categories before channel access, as shown in Table 5.10. In the CAP, each traffic category has different probabilities for getting a transmission chance and different times for retransmission. The traffic categories in Table 5.10 follow that of IEEE 802.1D [51]. A new emergency or medical event report is added. Table 5.11 defines transmission probability of different

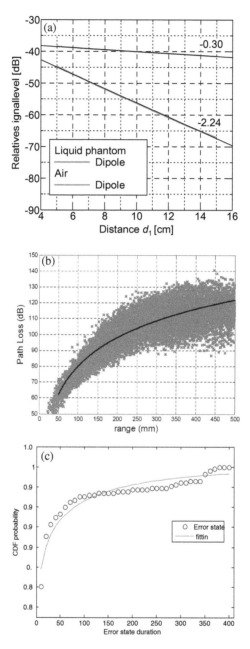

Fig. 5.24 (a) Measured path loss of MICS signal in [52], (b) simulated path loss of MICS signal in [53], and (c) measured fading period due to movement in the UWB band.

Table 5.10. User Priority Mapping.

Priority	User Priority	Traffic designation	Frame type
Lowest	0	Background (BK)	Data
	1	Best effort (BE)	Data
	2	Excellent effort (EE)	Data
	3	Controlled load (CL)	Data
Highest	4	Video (VI)	Data
	5	Voice (VO)	Data
	6	Network control	Management
	7	Emergency or medical event report	Data

Table 5.11. Contention Window Bounds for CSMA-CA and Contention Proba-bility Threshold for Slotted ALOHA.

User priority	CSMA-CA		Slotted Aloha	
	CWmin	CWmax	CPmax	CPmin
0	16	64	1/8	1/16
1	16	32	1/8	3/32
2	8	32	1/4	3/32
3	8	16	1/4	1/8
4	4	16	3/8	1/8
5	4	8	3/8	3/16
6	2	8	1/2	3/16
7	1	4	1	1/4

traffic categories (the CSMA will be discussed in Section 5.5.6). This provides different QoS for different traffics.

5.5.2.2   Contention-free period

The CFP is immediately after the CAP and will complete before the next beacon. The CFP consists of multiple guaranteed time slot (GTS). The CFP length is configurable with minimum length equal to zero.

Communications in the CFP is reservation based. The allocation of GTS is controlled by the coordinator to guarantee the required QoS and reliability for specific traffics. A WBAN node may inform the coordinator its QoS requirement during association. Multiple contiguous GTS can be granted to one device. Allocation of GTS can also be obtained through poll command.

The communication shall start only at the beginning of the GTS allocation. And it must complete before the end of the last allocated GTS. The access of GTS for a device can be periodic or on-demand. The periodic GTS access means the device is grated the same GTS allocation in every M ($1 \leq M < \infty$) superframes. It is designed for scheduled traffics, like medical waveforms and voices which occur periodically. After a GTS reservation is granted, the device can always use them until the reservation expires. The on-demand GTS access is one-time GTS allocation, which can be in the same superframe or the next superframe. It is designed for unscheduled traffic, like network managements, retransmissions, or low duty cycle events. As discussed in Section 5.3, the on-body WBAN channels is dominated by short-term fading channels, an intra-superframe retransmissions can greatly decrease traffic latency.

5.5.2.3   Poll command in contention-free period

Poll command is used by coordinator to inquire nodes whether there is any pending traffic to send or it is ready for reception. A particular reason for having poll command is the MICS system. According to the FCC's MICS rule, the implant device works in a reactive manner for except medical events. This implies that implant devices must be inquired first, and then answer the inquiry. It is, therefore, in the grey area whether the beacon is an "inquiry" or not. Although beacon is generally considered as a broadcast message, it does have mandatory CAP to allow devices to transmit. The poll command bypasses the grey area by specifying a two-way transmission session. The biggest difference between poll and beacon is that the poll is specified with a destination address, while the beacon is not. In other words, the poll command particularly inquires if particular devices have data to send. Therefore, using the poll complies with the FCC regulation for MICS systems.

The poll command is controlled by coordinator and it may appear regularly or on-demand in the unoccupied GTSs which have not allocated to any device. The poll command can address to a single device by a unique address or any device by the public address. The former allows an improvised access of GTS for a device. This can be used for retransmission or additional traffics beyond the capacity of allocated GTS. As illustrated in Figure 5.25, a poll command defines a communication period which can be immediately after the poll command, or after a while of the poll command. In Figure 5.25, an explicit poll command is used to enquire pending data. The explicit poll may emit regularly from the coordinator in unallocated GTSs based on a predefined parameter. The implicit poll command combines poll command and

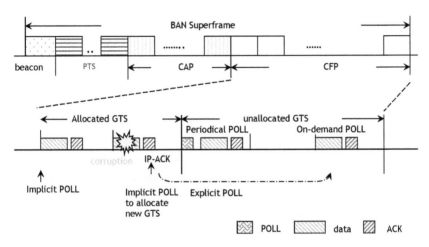

Fig. 5.25 Poll command in CFP.

acknowledgement (ACK) to indicate transmission corruptions, and requesting a retransmission. In such a case, it is referred as the on-demand poll.

5.5.2.4 Priority access period

The PAP is immediately after the beacon in the superframe. The PAP is dedicated for the emergency/alarm message, which is the highest priority defined in Table 5.10. Any other categories of traffics shall not take the PAP. The PAP is comprised by priority time slot (PTS), which is a physical time period immediately after the beacon, and embedded PAP, which are the unoccupied slots in both CAP and CFP. If transmission in the PTS fails, the highest priority traffic can use embedded PAP to contend the channel. It is not necessary to wait until the next superframe. The emergency/alarm message can be either uplink or downlink. Although emergency/alarm messages are typically low duty cycle, more than one medical sensor may simultaneously detect abnormal signals in the life-critical situation. Once the emergency traffic occurs, devices shall listen to the beacon to know the configuration of current superframe. If a GTS has been allocated in the current superframe, the device can transmit the emergency traffic in both PAP and the allocated GTSs. The channel access in PAP is contention based because it is in-efficient to allocate fixed radio resources to the low duty cycle nature of the emergency traffic.

The duration of PTS is configurable. For medical applications, it usually spans for two or three slots. For non-medical applications, it can be set to 0. The PTS in the superframe provides guaranteed channel access for emergency traffic, which is required by the FCC.

The PTS can be used for both uplink and downlink emergency communications. For downlink emergency traffics, the coordinator will indicate the addressed device in the beacon. The coordinator transmits emergency traffic in the scheduled communication window, and waits for the ACK from the node. For uplink emergency traffics, CAP described in Section 5.5.2.1 is adopted. If a data transmission failed, the coordinator may use on-demand poll command for re-transmission in the later embedded PAP. Although the channel access for uplink emergency traffic is contention cased, high probability of packet collision is not expected due to the low duty cycle nature of emergency traffics.

Although the PTS in the superframe provides the guaranteed channel access for emergency traffic, it is wasteful in bandwidth as the emergency priority traffic is usually low duty cycle. The embedded PAP corresponds to the unoccupied slots in both CAP and CFP. In other words, the embedded PAP partially overlaps with the CAP and the CFP. It provides a second opportunity for emergency traffics which failed to be transmitted in the PTS.

5.5.3   Non-Beacon Mode

When the coordinator does not regularly emit beacons, the WBAN piconet is said to operate under the non-beacon mode [49]. This is a power saving mode for devices which cannot or do not want to listen to beacons. For very low duty cycle applications, periodical wakeup and listening to the beacon for synchronization may consume more power than what that used for traffic payload. The non-beacon mode can also be used for very simple networks which have only a few (1 or 2) nodes. The beacon mode is overloaded for such applications.

One of key issue in MICS system is to wake up implant node per FCC's MICS rule, which must enter inactive state to save battery power. The MICS rule mandate that the implant node must work in the challenge–response mode except medical implant event report. But the implant nodes have no idea which channel the external coordinator will work on. The state-of-art solution depends on an out-band signaling to wake up implant and inform the working channel. For example, Zarlink uses OOK modulation in 2.4 GHz ISM band to wake up an implant node [50]. Some implant is wake up by magnet. The wakeup circuit at the implant can therefore be very simple, even without crystal clock. However, the out-bound signalings are out of the scope of IEEE 802 specifications.

In the non-beacon mode, there is no concept of superframe structure. Communications in the non-beacon mode are divided into handshake session and communication session with fixed durations. The handshake session consists of ALOHA slots. The poll command is used in the handshake session to confirm that the counterpart is ready for reception. The devices and the coordinator shall wake up at the rendezvous time and start the handover session. The deviation of the clocks may be very large after long time of sleeping. Comparing with the data frame, the poll command is short and therefore will not consume too much power even if it is lost. The communication session is uniquely occupied by one device which have successful handshake. The time reference is the start of first received poll.

As shown in Figure 5.26, either the coordinator or the implant device can initiate the handshake. They are used for downlink traffics and downlink traffics, respectively. The only difference is who speaks first. For the coordinator initiated handshake, the device shall enter reception state and cyclically tunes its receiver to each MICS channel and dwells for a while. After scanning the whole MICS channel and selecting one, the coordinator transmits a group polls with broadcast address or a unique address with a given separation. After each transmission, the coordinator listens for a packet arrival at the channel. If no T-poll command is heard, the implant node moves to the next MICS channel. The implant repeats the action until a poll is heard. Upon receiving a poll command, the device shall reply an ACK to indicate that it is ready for communications. The ACK frame may be corrupted due to collision or bad channel. It does not matter because the coordinator repeats T-poll transmission in the same channel. The coordinator may provide the implant nodes with an ordered list of channels to decrease likelihood when required to choose a new channel. This finishes the first handshake between external coordinator and implant node. Then both the coordinator and the implant device enter the communication session for traffic. If the handshake session failed, which could be due to bad channel conditions, clock offsets or packet collisions, the coordinator and the implant device shall exercise an exponential backoff,

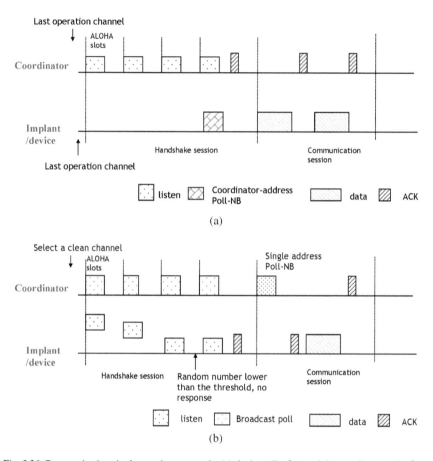

Fig. 5.26 Communications in the non-beacon mode: (a) device talks first and (b) coordinator talks first.

and make another attempt after the backoff period. If the handshake session expired, there will be no communication session. For MICS systems, implants should try to listen to different channels for poll command since they have no knowledge on which channel the coordinator will select. The MICS communication is organized into session which cannot be over 300 s. After each session, the implant node will enter low power mode. To initiate a new session, the coordinator can wake up the implant node by poll with a unique address. To minimize collision and save battery power, the coordinator will only talk with one implant in each session.

For the device initiated handshake, i.e., medical implant event in MICS system, the device will send polls first. Different from the coordinator initiated handshake, multiple devices may wake up simultaneously. The channel access mechanism is based on pure ALOAH for the same reasons described in Section 2.2.2. The packet collision contributes another reason for the hand-shake failure. If the handshake session failed, exponential backoff shall be conducted for the next opportunity.

5.5.4 Power Management

For medical applications, battery power consumption is a big concern. For low duty cycle applications, it is suggested to work in the non-beacon mode to reduce power for periodical clock synchronization. The coordinator can also save power by defining inactive superframe to skip beacon broadcasts when it does not expect any traffic from nodes and no need for piconet management. In the beacon mode, the nodes can shut down most part of its circuit. The coordinator and node must schedule the wakeup period before entering the low power mode.

5.5.5 Coexistence and Interference Mitigation

The coexistence issues in WBAN include the following:

- coexistence between medical application and non-medical appli-cation within a WBAN piconet.
- coexistence between medical WBAN piconet and non-medical WBAN piconet.
- coexistence among neighbor WBAN piconets.

The first issue is solved by the central control MAC protocol because both medical traffics and non-medical traffics are in the same piconet. The coordi-nator can control the allocated GTSs in CFP to medical traffics and to non-medical traffics. In the beacon mode, PAP is especially designed for life-critical medical emergency message; any other traffic cannot use it. The defined user priority gives medical applications higher priority than non-medical applica-tions in the CAP. Alternatively, the coordinator may create a medical super-frame and a non-medical superframe for each application.

For the second and the third issues, the most effective way is to allocate medical WBAN and non-medical WBAN in different frequency bands defined

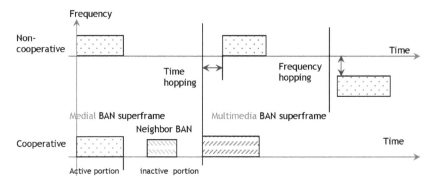

Fig. 5.27 Coexistence consideration of WBAN.

in Chapter 2. When a WBAN is created, the coordinator will scan the channel and select an unoccupied channel to operate in if possible. When multiple WBAN piconets work in the same frequency band, the solution depends on whether different WBAN piconets can cooperate with each other. In the cooperative manner, different WBAN piconets are synchronized with a clock. Then all WBAN piconets can arrange the active portion of the superframe in the inactive period of the other superframe. If they cannot cooperate with each other, each WBAN piconet can conduct time hopping and channel hopping for better coexistence as shown in Figure 5.27. However, a WBAN piconet shall not hop to a new channel in the middle of a superframe. The time hopping pattern and channel hopping pattern are managed by coordinator. Nodes can get them when the connection is established.

5.6   Summary of Chapter

In this chapter, we have presented MAC consideration and design for WBAN for medical applications. Section 5.2 focuses on the coexistence issues for multiple wireless technologies in the integral medical environment. The energy-based CCA is considered because the 2.4 GHz ISM band is too crowded to apply feature-based CCA for simple medical sensors. The asymmetric CCA issue has been found in both AWGN channel and fading channel. In the considered integrated medical environments, WPAN is oversensitive to 802.11b signals and WLAN is insensitive to 802.15.4 signals. There is an HECR, within which energy-based CCA can effectively avoid possible packet collisions.

However, when beyond the HECR, WLAN lose its sense to 802.15.4 signals. The asymmetric CCA puts WPAN traffic in a secondary position.

Section 5.3 investigates dynamic on-body fading from MAC point of view. By classifying the channel into good channels and bad channels per relative path loss in the dynamic action, statistics analysis reveals that the short durations are dominant in both good channels and bad channels. A 5-state Fritchman model has been presented to describe the transit between good channels and bad channels. Then the optimal backoff strategy for retransmission over the dynamic on-body has been studied. A preliminary analysis attributes the middle- and long-term fading to the intended movements and the dominant short-term fading to the unintentional movements.

Section 5.4 presents a general description of WBAN and MAC requirements for QoS, power consumption, and coexistence.

Section 5.5 finally provides a unified MAC design for WBAN which is independent of underlying PHYs including UWB, narrowband, and MICS for both medical and non-medical applications. The traffic is classified into different priorities before buffering in the communication queue. The beacon mode is mainly designed for dependable and guaranteed QoS for most important medical traffic and majority real-time traffic. A new PAP for life-critical message is introduced. The non-deacon mode is mainly designed for MICS system and low duty cycle applications. Power consumption is saved by inactive superframe and optional beacon listen.

References

[1] I. Demirkol, C. Ersoy, and F. Alagoz, "MAC protocols for wireless sensor networks: a survey," *IEEE Communications Magazine*, vol. 06, no. 4, pp. 115–121, 2006.

[2] A. Bachir, M. Dohler, T. Watteyne, and K. K. Leung, "MAC essentials for wireless sensor networks," *IEEE Communications Survey & Tutorials*, vol. 12, no. 2, pp. 222–248, 2010.

[3] O. Omeni, A. C. W. Wong, A. J. Burdett, and C. Toumazou, "Energy efficient medium access protocol for wireless body area sensor networks," *IEEE Transactions on Biomedical Circuits and Systems*, vol. 2, no. 4, pp. 251–259, 2008.

[4] B. Otal, L. Alonso, and C. Verikoukis, "Highly reliable energy-saving MAC for wireless body sensor networks in healthcare systems," *IEEE Jounal on Selected Areas in Communications*, vol. 27, no. 4, pp. 553–565, 2009.

[5] S. J. Marinkovic, E. M. Popovoco, C. Spagnol, S. Paul, and W. P. Marnane, "Energy-efficient low duty cycle MAC protocol for wireless body area networks," *IEEE Transactions on Information Technology in Biomedicine*, vol. 13, no. 6, pp. 915–925, 2009.

[6] R. Istepanian, E. Jovano, and Y. Zhang, "Guest editorial introduction to the special section on M-Health: Beyond seamless mobility and global wireless healthcare connectivity,"

IEEE Transactions on Information Technology in Biomedicine, vol. 8, no. 4, pp. 405–414, 2004.

[7]   A. Soomro and D. Cavalcanti, "Opportunities and challenges in using WPAN and WLAN technologies in medical environments," *IEEE Communication Magazine*, vol.45, no.2, pp. 114–122, 2007.

[8]   D. Cypher, N. Chevrollier, N. Montavont, and N. Golmie, "Prevailing over wires in health-care environments: Benefits and challenges," *IEEE Communication Magazine*, vol. 44, no. 4, pp. 56–63, 2006.

[9]   A. Lymberis, "Smart wearable systems for personalized health management: current R&D and future challenges," *International Conference of the IEEE*, vol. 4, pp. 3716–3719, 2003.

[10] N. Golmie, D. Cypher, and O. Rebala, "Performance analysis of low rate wireless technologies for medical application," *Computer Communication*, vol. 28, no. 10, pp. 1255–1275, 2005.

[11] N. Golmie, N. Chevrollier, and O. Rebala, "Bluetooth and WLAN coexistence: Challenges and solutions," *IEEE Wireless Communication Magazine*, vol. 10, no. 6, pp. 22–29, 2003.

[12] I. Howitt and J. A. Gutierrez, "IEEE 802.15.4 low rate — wireless personal area network coexistence issues," IEEE *Wireless Communications & Networking Conference*, vol. 3, pp. 1481–1486, 2003.

[13] S. Pollin, M. Ergen, and A. Dejonghe, "Distributed cognitive coexistence of 802.15.4 with 802.11," *International Conference on Cognitive Radio Oriented Wireless Networks and Communications*, pp. 1–5, 2006.

[14] Y. Xiao and N. H. Vaidya, "On physical carrier sensing in wireless ad hoc networks," *IEEE International Conference on Computer Communications*, vol. 4, pp. 2525–2535, 2005.

[15] J. Zhu, S. Roy, X. Guo, and W. S. Conner, "Leveraging spatial reuse in 802.11 mesh networks with enhanced physical carrier sensing," *IEEE International Conference on Communications*, vol. 7, pp. 4004–4011, 2004.

[16] H. Zhai and Y. Fang, "Physical carrier sensing and spatial reuse in multirate and multihop ad hoc network," *IEEE International Conference on Computer Communications*, pp. 276–285, 2006.

[17] I. Ramachandran and S. Roy, "On the impact of clear channel assessment on MAC performance," *IEEE GLOBECOM*, pp. 1–5, 2006.

[18] S. M. Kay, *"Fundamentals of Statistical Signal Processing: Detection Theory,"* Prentice-Hall PTR, New Jersey, 1998.

[19] V. Erceg, *et al.*, "An empirically based path loss model for wireless channels in suburban environments," *IEEE Journal on Selected Areas in Communications*, vol. 17, no. 7, pp. 1205–1211, 1999.

[20] A. Sonneschein and P. M. Fishman, "Radiometric detection of spread-spectrum signal in noise," *IEEE Transactions on Aerospace Electronic Systems*, vol. 28, no. 3, pp. 654–660, 1992.

[21] P. S. Hall, Y. Hao, and Y. I. Nechayev, et al., "Antenna and propagation for on-body communication systems," *IEEE Antennas and Propagation Magazine*, vol. 49, no. 3, pp. 41–58, 2007.

[22] D. Miniutti, L. Hanlen, D. Smith, and A. Zhang, *et al.*, "Dynamic narrowband channel measurements around 2.4 GHz for body area networks" IEEE P802.15-08-0033-00-0006, 2008.

[23] S. L. Cotton and W. G. Scanlon, "Characterization and modeling of the indoor radio channel at 868MHz for a mobile bodyworn wireless personal area network," *IEEE Antennas and Wireless Propagation Letters*, vol. 6, pp. 51–55, 2007.

[24] D. Cypher, N. Chevrollier, N. Montavont, and N. Golmie, "Prevailing over wires in healthcare environments: Benefits and challenges," *IEEE Communication Magazine*, vol. 44, no. 4, pp. 56–63, 2006.

[25] P. Sadeghi, R. A. Kennedy, and P. B. Rapajic, et al., "Finite-state Markov modeling of fading channels," *IEEE Signal Processing Magazine*, vol. 25, no. 5, pp. 57–80, 2008.

[26] E. N. Gilbert, "Capacity of burst-noise channel," *Bell System Technical Journal*, vol. 39, pp. 1253–1266, 1960.

[27] E. O. Elliott, "Estimates of error rates for codes on burst-noise channels," *Bell System Technical Journal*, vol. 42, pp. 1977–1997, 1963.

[28] B. D. Fritchman, "A binary channel characterization using partitioned Markov chains,"*IEEE Tranactions on Information Theory*, vol. 13, no. 2, pp. 221–227, 1967.

[29] H. S. Wang and N. Moayeri, "Finite-state Markov channels — a useful model for radio communications channels," *IEEE Transactions on Vehicular Technology*, vol. 44, no. 1, pp. 163–171, 1995.

[30] J. Garcia-Frias and P. M. Crespo, "Hidden Markov models for burst error characterization in indoor radio channels," *Transactions on Vehicular Technology*, vol. 46, no. 4, pp. 1006–1020, 1997.

[31] M. Kim, J. Takada, B. Zhen, L. Materum, T. Kan, et al., "Statistical Property of Dynamic BAN Channel Gain at 4.5GHz," 802.15-08-0489-02-0006.

[32] Bin Zhen, Minseok Kim, Jun-ichi Takada, and Ryuji Kohno, "Finite-state Markov model for on-body channels with human movements," *IEEE Conference on Communications*, May 2010.

[33] L. E. Baum, T. Petrie, G. Soules, and N. Weiss, "A maximization techniques occurring in the statistical analysis of probability functions of Markov chain," *Annuals of Mathematical Statistics*, vol. 41, no. 1, pp. 164–171, 1970.

[34] A. A. Serra, P. Nepa, G. Manara, and P. S. Hall, "Diversity for body area networks," URSI General Assembly, 2008.

[35] E. J. Podlaha and H. Y. Cheh, "Modeling of cylindrical alkaline cells. VI: Variable discharge conditions," *Journal of the Electrochemical Society*, vol. 141, no. 1, pp. 28–35, 1991.

[36] M. Zorzi and R. R. Rao, "Energy constrained error control for wireless channels," *IEEE Personal Communications*, vol. 4, no. 6, pp. 27–33, 1997.

[37] M. Zorzi and R. R. Rao, "Perspectives on the impact of error statistics on protocols for wireless networks," *IEEE Personal Communications*, vol. 6, no. 5, pp. 32–40, 1999.

[38] K. Y. Yazdandoost, "Channel Model for Body Area Networks (BAN)," IEEE 802.15-08-0033-07, 2008.

[39] B. Zhen, M. Kim, J. Takada, and R. Kohno, "Characterization and modeling of dynamic on-body propagation at 4.5 GHz," *IEEE Antennas and Wireless Propagation Letters*, vol. 8, pp. 1263–1267, 2009.

[40] M. Kim and J. Takada, "Statistical model for 4.5 GHz narrowband on-body propagation channel with specific actions," *IEEE Antennas and Wireless Propagation Letters*, vol. 8, pp. 1250–1254, 2009.

[41] J. (Andrew) Zhang, D. Smith, L. Hanlen, D. Miniutti, D. Rodda, and B. Gilbert, "Stability of narrowband dynamic body area channel", *IEEE Antenna and Wireless Propagation Letters*, vol. 8, pp. 53–56, 2009.

[42] D. Smith, L. Hanlen, J. (Andrew) Zhang, D. Miniutti, David Rodda, and B. Gilbert, "Characterization of the dynamic narrowband on-body to off-body area channel," *IEEE Communication Conference*, 2009.

[43] D. Lewis, "BAN use cases summary," *IEEE 15-08-0407-00-0006*.

[44] L. Schwiebert, S. K. S. Gupta, and J. Weinmann, "Research challenges in wireless networks of biomedical sensors," *ACM Conference on Mobile Computing and Networking*, pp. 151–165, 2001.

[45] IEEE P1073.0.1.1/D01J, "Draft guide for health informatics-point-of-care medical device communication-technical report-Guidelines for the use of RF wireless technology," 2006.

[46] B. Zhen, M. Patel, S. Lee, and E. Won, "Body area networks (BAN) technical requirements," *IEEE 802.15-07-0867-03*.

[47] IEEE standard 802.15.4b-2006.

[48] IEEE standard 802.15.4a-2007.

[49] Bin Zhen, Grace Sung, Huanbang Li, Ryuji Kohno, "NICT's MAC proposal to IEEE 802.15.6- document," *IEEE 802. 15-09-0814-02-0006*.

[50] http://www.zarlink.com/zarlink/hs/82_ZL70101.htm.

[51] IEEE 802.1D, "IEEE Standard for local and metropolitan area networks: Media Access Control (MAC) Bridges," 2004.

[52] T. Aoyagi, J. Takada, K. Takizawa, N. Katayama, T. Kobayashi, et al., "Channel modes for wearable and implant WBANs," *IEEE 802.15-08-0416-01-0006*.

[53] K. Sayrafian, K. Yekeh Yazdandoost, J. Hagedorn, J. Terril, W. Yang, and R. Kohno, "A Statistical path loss model for MICS," *IEEE 802.15-08-0519-00-0006*.

6

Standardization

6.1 Introduction

Wireless communication is a great way of connecting users and sharing data with each other. There are many companies that provide different hardware devices and software applications. Without coordination among products from different companies, there can be chaos, unmanaged communications, and disturbance. Therefore, it is necessary to make common rules that all manufacturers should adopt and produce products based on to enable interconnection and coexistence. Various standards are developed for the purpose.

Wireless communication by characterization is involved in radio wave transmissions. The radio spectrum resource is very much required by different users such as radio and TV stations, cellular phones, laptops, satellite links, military, and so on. Therefore, to develop any wireless telecommunication system, it is thus compulsory to obey the frequency regulation made by International Telecommunication Union (ITU) and local authorities. As a result, each wireless standard must specify the objective frequency bands and emission power level in accordance with corresponding radio regulations.

Moreover, specifications of medium access control (MAC) protocols and physical (PHY) layer parameters to be implemented in dedicated devices must be defined in order to efficiently use the specified frequency band without interfering with other devices. A new standard must have its own uniqueness compared with already existing standards. The specifications of either PHY or MAC of a new standard should ensure cooperation and do not interfere with existing standards. A good standard is also expected to be of huge market penetrating potentiality as well as to be of great technical feasibility and economic feasibility.

Among different organizations that are developing wireless standards, the Task Group 15.6 (TG6) was authorized by the IEEE 802 Local and Metropolitan Area Network Standards Committee with the charter of drafting a wireless standard for WBAN [1]. TG6 is one of the in-operation task groups of Work Group 15 (WG15) for WPAN. The latter is targeting at personal wireless services with a communication distance of up to several 10 m. WG15 has developed several famous wireless standards including PHY for Bluetooth (802.15.1-2005), PHY and MAC for ZigBee (802.15.4-2003, 2006). Moreover, IEEE 802.15.4a-2007 defines PHY using UWB to support low-rate service with ranging capability. Some recently established standards include 802.15.4c-2009 and 802.15.4d-2009 which are modified PHYs of 802.15.4-2006 to fit Chinese and Japanese regulations, 802.15.5-2009 which deals with MAC and up layer control for mesh network, and 802.15.3c-2009 which defines high-data-rate PHY on millimeter wave band.

Other international standardization organizations working on PHY and MAC specifications for wireless networks include International Organization for Standardization (ISO), International Electrotechnical Commission (IEC), European Computer Manufacture Association (ECMA), and so on. There are also a number of industrial collaborated consortiums that define specifications of products, such as Bluetooth Special Interest Group (Blutooth SIG), ZigBee Alliance, and Continua Health Alliance. Among all these organizations, TG6 within the IEEE 802.15 WG is the first group focusing on WBAN. Continua Health Alliance focus on healthcare services and look at various technologies including Internet and end devices. Continua Health Alliance announced their Bluetooth product for physiological monitoring solutions in 2008 [2]. The proposed solutions adopted ISO/IEEE 11073 data protocol and aimed to provide interoperable standards in telemedicine and chronic disease management. ZigBee Alliance also expressed its desire to provide healthcare solutions by announcing collaboration with American Telemedicine Association (ATA) [3]. For medical and healthcare purpose, there is a standardization body called IEEE 11073, which defines medical device communication. Moreover, there is an eHealth project within ETSI concentrating on ICT for healthcare and medical purposes. In the rest of this chapter, we overview the activity of TG6 of IEEE 802.15, IEEE 11073, and ETSI eHealth project.

6.2 IEEE 802.15.6: Body Area Network

6.2.1 WBAN Defined at TG 15.6

After successful launching of a number of WPAN standards, WG 15 of IEEE 802 reached to a point to define new directions and new projects. Consequently, a Standing Committee of Wireless Next Generation (WNG SC) was set up in January 2006. The WNG SC collected a list of candidate technologies from the WG 15 members. Some of them are shown below [4].

- Extremely low power and energy efficient radio.
- Body area network.
- MIMO technology for WPAN.
- Software-defined radio.
- Mobile, nomadic mesh network.
- Heterogeneous WPAN interoperability and internetworking.
- Wireless system coexistence and dynamic mitigation techniques.

Among these technologies, BAN attracted a large number of attentions from the participating members and proposals of starting a BAN group came out. As a response to the proposals, an interest group of BAN (IG-BAN) was formed in May 2006. In November 2006, the IEEE 802 Executive Committee (EC) formally approved the group as a study group. Furthermore, In December 2007, the group was approved as a task group.

A complete definition of BAN given by TG6 is as follows [5]:

> *This is a standard for short range, wireless communication in the vicinity of, or inside, a human body (but not limited to humans). It can use existing ISM bands as well as frequency bands approved by national medical and/or regulatory authorities. Support for quality of service (QoS), extremely low power, and data rates up to 10 Mbps is required while simultaneously complying with strict non-interference guidelines where needed. This standard considers effects on portable antennas due to the presence of a person (varying with male, female, skinny, heavy, etc.), radiation pattern shaping to minimize SAR (specific absorption rate) into the body, and changes in characteristics as a result of the user motions.*

6.2.2 Technical Requirements and Possible Solutions

6.2.2.1 Main Technical Requirements

TG6 looks at a large scope of applications including not only medical and healthcare services but also consumer-centric electronics. As a result, technical requirements for WBAN also present varieties. The main technical requirement parameters summarized by TG6 are as follows:

- Number of devices: six nodes for a typical medical WBAN but should be efficiently scalable up to 256 nodes.
- Data rate: 10 kbps–10 Mbps.
- Transmission range: at least 3 m.
- Packet error rate: less than or equal to 10% for a 256 octet payload with a link success probability of 95%.
- Security: authentication, data integrity, and encryption operations should be supported.
- Quality of service (QoS): PHY and MAC need to provide and guarantee latency control, priority control, power saving, support for different traffics.
- Simultaneously operated WBANs: MAC and PHY should support co-located operation of at least 10 randomly distributed WBANs in a volume of $6 \times 6 \times 6$ m.
- Coexistence: coexistence between WBAN and other wireless devices shall be addressed.

It should be noted that the technical requirements provide a bottom line. Possible solutions need to demonstrate how the technical requirements can be met. In that sense, technical requirements can be regarded as the criteria to evaluate possible solutions. However, because of the wide range of technical parameters, it would be difficult to define a single efficient solution. Consequently, TG6 members came to the consensus that the standard should accept multiple PHYs, allow partial solutions for a single PHY, and define a single MAC.

Multiple PHY solutions mean that the standard should be based on different radios on different frequency bands. The reason behind is the different frequency bands available. It would be more efficient to define radios in accordance with the regulations on the objective frequency bands.

The reason for allowing partial solutions is that it is difficult for a single solution to meet all technical requirements with reasonable complexity and form factor. Therefore, solutions that partly satisfy technical requirements should be acceptable as a part of the standard. However, as a standard, IEEE 802.15.6 must present unity. Then, a single common MAC is important to provide a common platform between multiple PHYs and enable unified control and interconnection.

6.2.2.2 Main PHY Parameters

Main PHY proposals to TG6 include narrowband PHY, UWB PHY, and PHY using human body communication (HBC). The main features of the three PHYs are summarized as follows.

(1) Narrow band PHY

- Frequency band
  - MICS: 402–405 MHz
  - WMTS: 420–450 MHz, 868/915/950 MHz
  - ISM and plus: 2.4–2.48 GHz and 2.36–2.4 GHz
- Modulations
  - Basic mode: $\pi/2$-DBPSK and $\pi/4$-DQPSK
  - Optional mode for Japanese WMTS bands: GMSK
  - Optional mode: $\pi/8$-D8PSK, GFSK
- Error correction codes
  - BCH (63, 51) for data
  - BCH (31, 19) for header
- Data rates
  - 50 kbps–1 Mbps

(2) UWB PHY

- UWB frequency bands
  - UWB high band: 7.25–8.5 GHz
  - UWB low band: 3.1–4.8 GHz,

- Modulations
  - IR-UWB (non-coherent detection)
  - IR-UWB (differentially coherent detection)
  - FM-UWB (frequency modulated UWB)
- High QoS mode
  - Hybrid Type-II automatic repeat-request (ARQ)
- Data rates
  - 0.5–10 Mbps

(3) HBC PHY

- HBC frequency bands
  - 10–50 MHz
- Transmission method
  - Frequency-Selective Digital Transmission
- Data rate
  - 125 kbps–2 Mbps

6.2.2.3 Main MAC Features

TG6 defines both a beacon mode and a non-beacon mode. Beacon mode is based on a superframe structure shown in Figure 6.1. It can be seen that there are two copies of the same structures, such as two copies of random access phase (RAP) and two copies of exclusive access phase (EAP). The second RAP and EAP are introduced to reduce the latency for device to join a WBAN and guarantee the minimum latency of the highest priority traffic.

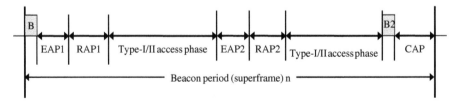

Fig. 6.1 Superframe structure in beacon mode.

The basic mechanism of channel access in EAP, RAP, and contention access phase (CAP) is either slotted ALOHA or slotted CSMA-CA depending on the underlying PHY. For narrowband PHY, it is suggested to adopt CSMA-CA. For UWB PHY, slotted ALOHA is preferred.

The coordinator can also send poll to a node in the unallocated contention-free slots. This is termed unscheduled access which can help to reduce latency of multi-medium traffic on the expense of power consumption. Two types of access phases are defined. However, only one of them is allowed in the superframe, which is indicated in the beacon. The major difference lies in the polled allocation. Type-I polled allocation is defined in terms of maximal time duration in which the polled node may use for its frame transactions in the allocation. Type-II polled allocation is defined in terms of maximal number of frames in which the polled node may transmit in the allocation. In other words, the duration of Type-I polled allocation is fixed, while the duration of Type-II polled allocation is variable, which may depend on the channel quality.

B2 packet is locally broadcasted by a coordinator to announce the beginning of CAP and provides group acknowledgment and coexistence information. Different from the beacon, B2 is not in charge of superframe structure before it. It only defines the duration of CAP and coexistence scheme.

6.2.3 Standardization

By May 2009, a total of 34 proposals were presented to TG6. After negotiation on several consecutive meetings, an agreement was reached to merge all 34 proposals together in March 2010. According to the agreement, TG6 members worked out the first draft of IEEE 802.15.6 in May 2010.

At the time of writing this book, TG6 gained 76.7% YES votes in the first letter ballot in July 2010. However, IEEE 802 requires the responsible TG to respond to all submitted comments in a process called comment resolution. The draft will be updated during the comment resolution procedure.

6.3 IEEE 11073: Point-of-Care Medical Device Communication Standards Committee

The IEEE 11073 dated back to 1982 when researchers recognized the benefits that plug-and-play features would bring to medical devices used in hospital, clinic, and operating rooms. The IEEE 11073 is recently renamed as ISO/IEEE 11073. The primary goal of this standard is to provide real-time plug-and-play interoperability for patient connected medical devices and facilitate the efficient exchange of vital signs and medical device data, acquired at the point-of-care, in all healthcare environments [11].

The ISO/IEEE 11073 has been achieved with a goal of implementing the standards in medical and healthcare facility. However, home healthcare is still missing. The IEEE 802 standard on WBAN is aiming to provide a standard for medical healthcare facilities as well as home-based one.

The ISO/IEEE 11073 family is based on an object-oriented system management paradigm. Data is modeled in the form of information objects that can be accessed and manipulated using an object access service protocol [10].

The ISO/IEEE 11073 standard family defines:

- How to structure data to be transmitted (DIM, domain information model).
- What commands are to be used to access this data (SM, service model).
- Communication states and what is to be sent or requested in each state (CM, communication model).

to exchange and evaluate vital signs data between different medical devices, as well as remote to control these devices.

The ISO/IEEE 11073 standards family are the following:

- 11073-00101 Health informatics–PoC medical device communication–Part 00101: Guide – Guidelines for the use of RF wireless technology.
- 11073-10101: 2004(E) Health informatics–Point-of-care medical device communication–Part 10101: Nomenclature.
- 11073-10201: 2004(E) Health informatics–Point-of-care medical device communication–Part 10201: Domain information model.
- 11073-20101: 2004(E) Health informatics–Point-of-care medical device communication–Part 20101: Application profile–Base standard.
- 11073-30200-2004 Health informatics–Point-of-care medical device communication–Part 30200: Transport profile–Cable connected.

- 11073-30300: 2004(E) Health informatics–Point-of-care medical device communication–Part 30300: Transport profile–infrared wireless.
- 11073-10404-2008 Health informatics — Personal health device communication–Part 10404: Device specialization — Pulse oximeter.
- 11073-10408-2008 Health informatics — Personal health device communication–Part 10408: Device specialization — Thermometer.
- 11073-10415-2008 Health informatics — Personal health device communication–Part 10415: Device specialization — Weighing scale.
- 11073-10417-2008 Health informatics — Personal health device communication — Part 10417: Device specialization — glucose meter.
- 11073-10441-2008 Health informatics — Personal health device communication — Part 10441: Device specialization — Cardiovascular fitness and activity.
- 11073-10442-2008 Health informatics — Personal health device communication — Part 10442: Device specialization — Strength fitness equipment.
- 11073-10471-2008 Health informatics–Personal health device communication–Part 10471: Device specialization–independent living activity hub.
- 11073-20601-2008 Health informatics–Personal health device communication–Part 20601: Application profile — Optimized Exchange Protocol.
- 11073-10404-2010 Health informatics–Personal health device communication–Part 10404: Device specialization–Pulse oximeter.
- 11073-10407-2010 Health informatics–Personal health device communication–Part 10407: Device specialization–Blood pressure monitor.
- 11073-10408: 2010 Health informatics–Personal health device communication–Part 10408: Device specialization–Thermometer.
- 11073-10415: 2010 Health informatics–Personal health device communication–Part 10415: Device specialization–Weighing scale.
- 11073-10417: 2010 Health informatics–Personal health device communication–Part 10417: Device specialization–Glucose meter.

- 11073-10471: 2010 Health informatics–Personal health device communication–Part 10471: Device specialization–Independent living activity hub.
- 11073-20601: 2010 Health informatics–Personal health device communication–Part 20601: Application profile–Optimized exchange protocol.

6.4   ETSI

6.4.1   eHealth Project

In Europe, the European Telecommunication Standards Institute (ETSI) Project eHealth (EP eHealth) is responsible for collecting and defining the health ICT-related requirements to input the requirements to the concerned ETSI Technical Bodies. The ultimate aim is that eHealth systems should offer users better access to more cost-effective healthcare services, irrespective of location.

In February 2009, EP eHealth published a technical report, which analyses user services models, technologies and applications supporting eHealth [8]. In the context, it provides a guide to future standardization may on occasion where future standards are required. Figure 6.2 describes the requirements for ubiquity, security, privacy, and reliability across the eHealth system and the supporting ICT technologies. From network point of view, the specific eHealth support networks include the following:

- BAN communication which connects implanted/wearable devices with a proxy device (typically the gateway) that connects the BAN with external infrastructures (PAN, Internet, etc.).
- PAN communications among eHealth devices around the person allowing transparent, secure, and trusty connection. This set of devices and nodes creates an eHealth environment with both intra- and inter-PAN communication capabilities.
- Personal network (PN) extends that PAN with other remote eHealth devices and services farther away using local wired or wireless connections as well as infrastructure-based connections and even multi-hop ad hoc networks to connect geographically dispersed personal nodes.

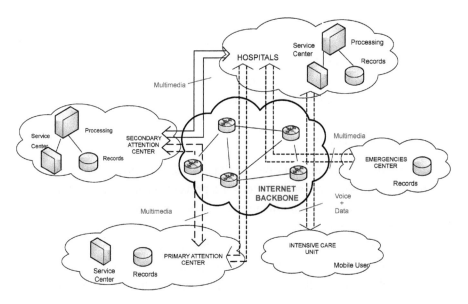

Fig. 6.2 The eHealth system illustrating links and users [8].

- PN-Federation is the aggregation of several cooperated PNs in order to share personal resources, services, and content which have been agreed by the owner previously to achieve a common objective that would not be possible by a single PN.

The potential radio technologies include RFID, WiFi, WiMax, UWB, GSM, UMTS, and satellite, while the fixed network technologies include IPv4/IPv6, NGN, and IMS.

The technical report is an initial step in developing eHealth user service models to address interoperable solutions for healthcare data collection, transmission, storage, and interchange, all with appropriate security, privacy and reliability. Work continues on two complementary technical reports, one collecting together architectures and service models for eHealth and the other mapping use cases and services to telecommunication services. It will identify any standardization gaps and a set of use cases linked to standards.

EP eHealth is also working to define the service requirements for short-range radio communication and networking such as QoS, security, privacy, reliability, robustness, frequency bands, regulatory requirements, and power consumption. Work on the personalization of eHealth services which match the needs of every user, even when those needs may be different from those of the majority, by using eHealth user profiles is a joint project of EP eHealth and ETSI's Human Factors Technical Committee. The object is to demonstrate how services and devices within eHealth can be interoperable and personalized by users to meet the individual's needs.

6.4.2 ETSI's Security

ETSI's Security Algorithms Group of Experts (SAGE) provide cryptographic algorithms and protocols specific to fraud prevention, unauthorized access to public and private telecommunications networks, and user data privacy.

The ISO Technical Committee (TC) 215 works on the standardization of health informatics to allow for compatibility and interoperability between independent systems. The ISO TC 215 consists of several Working Groups (WGs), each dealing with an aspect of electronic health records (EHR).

- WG 1: Data structure.
- WG 2: Messaging and communications.
- WG 3: Health Concept Representation.
- WG 4: Security.
- WG 5: Health Cards ≫Transitioned to a Task Force on Health Cards.
- WG 6: Pharmacy and Medication.
- WG 7: Devices.
- WG 8: Business requirements for Electronic Health Records.

6.5 Security and Privacy Risk

Wireless technology gives many benefits but these may come with unanticipated risks. The patient's private medical information stored on medical devices such as pacemakers or implantable cardiac defibrillators could be extracted, and more severely, their devices reprogrammed without the patient's authorization or knowledge [6]. As a government initiative, the Health Insurance Portability and Accountability Act of 1996 (HIPAA) present regulation on privacy, integrity, and access control requirement for patient data [9].

Security typically involves protecting data and protecting the network from pinching or destruction. Protecting data ensures that it arrives at the destination as it was transmitted and that no other entity was able to receive and decode the data. Protecting the network ensures that external attacks do not adversely affect network performance [7].

Implantable medical device such as pacemakers and implantable cardiac defibrillators can save lives and greatly improve a patient's quality of life. But, as these devices begin to interoperate in vivo and to use wireless communications with outside the body, they need to address security and patient privacy. There has never been a reported case of a patient with a medical device being targeted by hackers.

When considering medical device security and privacy, it is important to draw distinctions between classes of devices that have different physical properties and healthcare goals.

The range of applications supported by WBAN technology is quite wide, from patient condition monitoring to control of body functions. Likewise, the degree to which the patient will need security and protection of privacy will also vary.

6.6 Summary of Chapter

The development processes of standardization in different organizations are structured in different approaches. The technology is developed by manufacturers and operators, however, to be able to use the newly developed technology, a formal standardization should be created in standardization organizations. Along with changes in technology and developments, there might also be improvements in the standardization process according to the requirements of the wireless environment.

References

[1] IEEE802.15.TG6 at http://www.ieee802.org/15/pub/TG6.html.

[2] http://continuahealthalliance.createsend1.com/t/r/e/irtyll/bjtltmy/.

[3] http://zigbee.org/imwp/idms/popups/pop_download.asp?contentID=16815.

[4] Erik Schylander, "15WNG Guidelines for New Work Items," IEEE 802.15-06-0002-01, Jan. 2006.

[5] Project Authorization Request (PAR) for P802.15.6, 15-07-0575-09-0ban-ban-draft-par-doc.rtf.

[6] K. Yekeh Yazdandoost and R. Kohno, "Health Care and Medical Implanted Communications," in *13th International Conference on BioMedical Engineering*, December 2008.

[7] S. D. Baker and D. H. Hoglund, "Medical-grade, mission-critical wireless network," *IEEE Engineering in Medicine and Biology Magazine*, vol. 27, no. 2, pp. 86–95, March–April 2008.

[8] ETSI TR 102 764, "eHEALTH: Architecture: Analysis of user service models, technologies and application supporting eHealth," v1.1.1, 2009.

[9] Health Insurance Portability and Accountability Act of 1996 (HIPAA), http://www.hhs.gov/ocr/privacy/.

[10] J. Yao and S. Warren, "Appling the ISO/IEEE 11073 Standards to Wearable Home Health Monitoring System," *Journal of Clinical Monitoring and Computing*, vol. 19, no. 6, pp. 427–436, 2005.

[11] Health Informatics — Point-of-Care Medical Device Communication–Domain Information Model: ISO/IEEE 11073 Committee, December 2004.

Index

RIVER PUBLISHERS SERIES IN INFORMATION SCIENCE AND TECHNOLOGY

Other books in this series:

Volume 1
Traffic and Performance Engineering for Heterogeneous Networks
Demetres D. Kouvatsos
February 2009
ISBN: 978-87-92329-16-5

Volume 2
Performance Modelling and Analysis of Heterogeneous Networks
Demetres D. Kouvatsos
February 2009
ISBN: 978-87-92329-18-9

Volume 3
Mobility Management and Quality-of-Service for Heterogeneous Networks
Demetres D. Kouvatsos
March 2009
ISBN: 978-87-92329-20-2

Volume 4
Aspects of Kolmogorov Complexity: The Physics of Information
Bradley S. Tice
September 2009
ISBN: 978-87-9239-26-4

Volume 5
Biomedical and Environmental Sensing
J.I. Agbinya, E. Biermann, Y. Hamam, F. Rocaries and S.K. Lal (Eds.)
November 2009
ISBN: 978-87-9239-28-8

Volume 6
Pattern Recognition and Machine Vision
Patrick Shen-Pei Wang (Ed.)
March 2010
ISBN: 978-87-92329-36-3

Volume 7
Stealing Time: Exploration in 24/7 Software Engineering Development
Zenon Chaczko, Ryszard Klempous and Jan Nikodem (Eds.)
2010
ISBN: 978-87-92329-42-4

For Product Safety Concerns and Information please contact our EU
representative GPSR@taylorandfrancis.com
Taylor & Francis Verlag GmbH, Kaufingerstraße 24, 80331 München, Germany

www.ingramcontent.com/pod-product-compliance
Ingram Content Group UK Ltd.
Pitfield, Milton Keynes, MK11 3LW, UK
UKHW021121180425
457613UK00005B/170